Detecting the Stochastic Gravitational-Wave Background

Detecting the Stochastic Gravitational-Wave Background

Carlo Nicola Colacino

Ministero dell'Istruzione, dell'Università e della Ricerca, Italy

Morgan & Claypool Publishers

Rights & Permissions
To obtain permission to re-use copyrighted material from Morgan & Claypool Publishers, please contact info@morganclaypool.com.

ISBN   978-1-6817-4082-9 (ebook)
ISBN   978-1-6817-4018-8 (print)
ISBN   978-1-6817-4210-6 (mobi)

DOI   10.1088/978-1-6817-4082-9

Version: 20171201

IOP Concise Physics
ISSN 2053-2571 (online)
ISSN 2054-7307 (print)

A Morgan & Claypool publication as part of IOP Concise Physics
Published by Morgan & Claypool Publishers, 1210 Fifth Avenue, Suite 250, San Rafael, CA, 94901, USA

IOP Publishing, Temple Circus, Temple Way, Bristol BS1 6HG, UK

To Camilla, with love and gratitude, and to our uncommon children: Theo, Piero, Noemi and Gabriele.

Contents

Preface

The stochastic gravitational-wave background (SGWB) is the most mysterious source of gravitational radiation. Its detection will be a milestone for science and could dramatically change our knowledge of the Universe. During the writing of this book, gravitational radiation was detected experimentally for the very first time: this detection, announced on 11 February 2016, was hailed as a fantastic achievement for the scientific community. More detections followed and on 16 October 2017 a joint gravitational wave electromagnetic observation was announced: the birth of gravitational wave astronomy has finally become reality. Gravitational wave cosmology could be the most important and most exciting chapter of observations. In this book we review briefly what the SGWB is, its most likely sources and the data analysis techniques required to claim a successful detection.

None of the material presented here is original, everything that is written in this book has already appeared in the literature. The author of this book was a member of the LIGO/Virgo Collaboration for more than ten years and he was involved in the Stochastic Upper Limit Group and contributed to some of the results quoted here. However the way this material is presented is hopefully original and reflects the background of the author, who is not an astrophysicist but rather a theoretical cosmologist with a strong background in high-energy physics.

Acknowledgements

I want to thank the Research Institute for Nuclear and Particle Physics of the Hungarian Academy of Sciences, for the time I have spent there as a visiting scientist since 2007, and the Institute of Physics of Belgrade.

Author biography

Carlo Nicola Colacino

Carlo Nicola Colacino studied theoretical physics at the Sapienza University of Rome, working alongside one of the most famous Italian astrophysicists, Remo Ruffini, and one of the most promising high-energy theoreticians, Massimo Bianchi. He then worked on his PhD in Cagliari, under the guidance of Massimo Bianchi and Mariano Cadoni. His research career saw him spend years in Hanover, as a scientist on the project GEO600, Birmingham and Budapest within the framework of the LIGO Scientific Collaboration (LSC) as well as Pisa, within the Virgo project, where he worked specifically on the stochastic gravitational-wave background, detector characterisation and data analysis. He is currently working for the Italian Ministry of Education and Research as a teacher and outreach scientist, bringing the excitement of physics to young children. His main passions are cosmology, high-energy physics, choir music, rugby and probability: because of this, he is one of the leading scientists of a project launched by the Italian Government to inform people, especially young people, about the risks of gambling and games of hazard.

Chapter 1

A brief history of gravitational radiation

1.1 Introduction

More than 100 years ago, in 1915, Einstein published his general theory of relativity (GR). This theory changed dramatically our understanding of the gravitational force, or, to say it in modern parlance, interaction. Gravitation was no longer seen as a force, but rather as the curvature of spacetime itself. Newton's theory of gravitation, that had withstood the test of time for almost 300 years, became just an approximation of a radically different theory. It must be stressed that, although many physicists struggled to understand the novelty of Einstein's theory and Einstein himself did not win the Nobel Prize in Physics for GR—he was awarded the prize in 1921 for his pioneering work on quantum mechanics and his explanation of the photoelectric effect and, precisely because of the hostile reception by the physics community to his GR, did not collect the prize himself—100 years thereafter, Nature has always said a sound and clear yes to all predictions of GR. No experiment so far has even planted the smallest seed of doubt concerning the validity of GR. GR—the most beautiful of the physical theories because of its elegant and consistent mathematical formulation and richness—ranks amongst the most experimentally confirmed scientific theories, a milestone in our understanding of the Universe, its evolution and its structure. GR predicts the existence of ripples of spacetime that propagate with the speed of light, the so-called gravitational waves (GWs). This specific prediction was verified for the first time at the end of the last century in an indirect way: two radioastronomers, Hulse and Taylor, observed for more than 20 years the rotation period of the binary system PSR1913+16, a system made up of two neutron stars rotating around their centre of mass, and noticed that the period of rotation changed with time. They attributed this change to the energy loss due to the emission of GWs and showed that their data were in perfect agreement, within the experimental error, with the predictions of GR, actually one of the best agreements ever found in science between theory and data. Hulse and Taylor were awarded the Nobel Prize in 1993. On 11 February 2016 the Laser

Interferometric Gravitational Observatory (LIGO) in the US and the European project Virgo announced the first direct detection of GWs, GW150914 [1]. This was an historical moment. Two massive black holes had merged and the event's gravitational-wave signal detected. The event that was announced on 11 February but had been recorded on the instruments on 14 September 2015, therefore 14 September has internationally become Gravitational-Wave Day. Three subsequent detections of the same kind of event have followed since: GW151226 [2], GW170104 [3] and GW170814 [4]. As a result of these detections the Physics Nobel Prize 2017 was awarded to three LIGO scientists: Kip S Thorne (Caltech), Rainer Weiss (MIT) and Barry C Barish (Caltech). Another breakthrough was about to come: on 16 October 2017 it was announced that thanks to triangulation made by the three detectors, a GW signal, GW170817, had been recorded and its source localised in the sky, which made it possible to observe the event with electromagnetic detectors as well. The conclusion was reached that the radiation had been produced by the merging not of black holes but of neutron stars. Multimessenger astronomy, i.e. combining GW and electromagnetic observations, was officially born on 17 August 2017, the day the signal was detected by the interferometers [5].

1.2 Sources of gravitational radiation

The gravitational radiation that was detected on 14 September and announced on 11 February was produced by the coalescence of a binary system made up of two supermassive black holes of roughly 36 and 29 solar masses, respectively. The signal was named **GW150914**, from the date it was observed. Such coalescences are described by scientists as a three-step process and gravitational radiation is produced in all three different phases: first, as the two objects rotate around one another there is the inward spiralling of the two compact objects; as they get closer and closer they reach the merging phase—mergers of compact objects are the most violent events in the history of the Universe after the Big Bang itself—and finally there is the ringdown of the single resulting black hole. The coalescence of a compact binary was the most likely candidate for the first detection and remains the main source of gravitational radiation in the eyes of scientists, both theoreticians and data-analysts alike. GR describes rather well the first and third stage, i.e. the spiralling and the ringdown. Theoretical waveforms are known for these two phases and can give information on a number of physical parameters such as the masses of the compact objects. The strong gravitational fields produced during the merging phase cannot be described too accurately by GR but can be tackled by using numerical relativity. Data analysis is greatly enhanced by the use of the matched filtering technique.

Hulse and Taylor's source was of a different kind: pulsars are stars that emit pulses at radio frequencies at very precise, regular intervals as they rotate. Again, the waveform is very well known from GR. It is just a sine wave in the Solar System barycentre, a coordinate system at rest with respect to the Sun. The orbital motion of the Earth around the Sun and the rotational motion of the Earth around its own axis modulate the frequency and this makes data analysis slightly more complicated than in the previous case of the compact binary coalescences.

Supernovae, or bursts, are the third kind of source, and perhaps the most difficult to detect, for two reasons. First, we have no precise waveform of the gravitational signal such events could produce, so we cannot rely on matched filtering. Furthermore, the effects of supernovae on our detectors might look remarkably similar to instrumental noise, so it is hard to tell them apart. Therefore, bursts are so hard to detect.

The stochastic gravitational-wave background is something entirely different. With the acronym SGWB we describe all the 'random' sources, random because they arise from a large number of 'unresolved', independent and uncorrelated events. We are not talking about quantum physics here: the individual events that make up the SGWB are perfectly deterministic. It is their unresolved superposition that produces a random signal, exactly as the superposition of many broadcasting stations in the same frequency band sums up to confusion noise on our radios. Such a signal can be treated and described only statistically. This type of background can be the result of processes that took place in the very early Universe, shortly after the Big Bang, or could have arisen more recently during structure formation. It is almost impossible to detect with present-day technology, nevertheless it is perhaps the most interesting source: from a theoretical point of view, as we have already mentioned, the cosmological background, the one produced immediately after formation, when the Universe was perhaps less than a picosecond old, could—if detected—shed light on all mysteries of cosmology as well as high-energy physics. Our knowledge about the Universe comes entirely from electromagnetic observations. Our most detailed view of the Universe stems from the cosmic microwave background radiation (CMBR), an isotropic electromagnetic radiation that decoupled from matter around 5×10^5 years after the Big Bang. We have no consistent and established quantum theory of the gravitational interaction, but reliable theoretical estimates seem to suggest that if current detectors reveal an SGWB component of cosmological origin, then it will carry with it a picture of the Universe as it was about 10^{-22} s after the Big Bang. That would be a tremendous leap forward in our knowledge and therefore scientists refer to the SGWB of cosmological origin as the 'Rosetta Stone' of cosmology and of particle physics. From a more practical point of view, we know that there is a stochastic component hidden in every data set we record with our detectors. When we look for any of the previously listed sources, we are really searching for a needle in a haystack. This is sometimes very rewarding, as in the case of the first detection, but it can also be very frustrating and is in any case computationally very intensive. On the other hand, as we will see in this book, any time two GW detectors operate and take data simultaneously, there will be an SGWB component in our data. This component might be very well hidden below the noise floor, but we know it is there and this knowledge lets us gain further insight into the SGWB.

We will start by recalling how GWs arise in GR, then calculate the radiation from a rotating binary source with given parameters. We will also rederive Hulse and Taylor's result for the 1913+16 binary pulsar. We will then proceed to characterise mathematically the SGWB, discussing cosmological as well as astrophysical sources. A large section will also be devoted to the data analysis problem for the SGWB.

1.3 Gravitational waves from general relativity

Einstein's GR is a very complex theory, where gravitation is seen not as a force but rather as the curvature of spacetime itself. The essence of GR can be summarised in the famous words by John Archibald Wheeler 'matter tells space how to curve, space tells matter how to move' (the sentence would be perfect, however, if the word 'space' were to be changed into 'spacetime'). The mathematical formulation of this idea is given by the GR field equations:

$$R_{\mu\nu} - \frac{1}{2}g_{\mu\nu}R = 8\pi G T_{\mu\nu} \tag{1.1}$$

$R_{\mu\nu}$ is the so-called Ricci tensor, obtained by contraction from the Riemann, or curvature, tensor $R_{\mu\nu} \equiv R^{\alpha}_{\ \mu\alpha\nu}$. R is the the Ricci scalar $R = g^{\mu\nu}R_{\mu\nu}$. These two quantities, well known in Riemannian geometry, are purely geometrical, they measure the intrinsic curvature of the spacetime. $T_{\mu\nu}$ is the covariant form of the energy–momentum tensor, i.e. a quantity related to matter. This distinction is, however, not so clear-cut, as can be seen by taking the trace of both members from which we can recast equation (1.1) as

$$R_{\mu\nu} = 8\pi G\left(T_{\mu\nu} - \frac{1}{2}g_{\mu\nu}T\right) \tag{1.2}$$

The equations are highly nonlinear in the ten unknowns $g_{\mu\nu}$.

GWs are solutions to these equations in the weak-field approximation: we need to expand Einstein's equations around the flat space Minkowski metric, therefore, we put

$$g_{\mu\nu} = \eta_{\mu\nu} + h_{\mu\nu} \tag{1.3}$$

with $|h_{\mu\nu}| \ll 1$ and expand equation (1.1) to linear order in $h_{\mu\nu}$, which we treat as a quantity which transforms as a tensor under Lorentz transformations. The resulting theory is called the *linearised theory*. To first order in $h_{\mu\nu}$ the Ricci tensor becomes

$$R^{(1)}_{\mu\nu} \sim \left(\partial_{\lambda}\Gamma^{\lambda}_{\mu\nu} - \partial_{\nu}\Gamma^{\lambda}_{\lambda\mu}\right) + \bigcirc(h^2) \tag{1.4}$$

where the Christoffel symbol, as it is called in the framework of GR, or the affine connection, as it is known by differential geometry mathematicians is

$$\Gamma^{\lambda}_{\mu\nu} = \frac{1}{2}\eta^{\lambda\alpha}(\partial_{\mu}h_{\alpha\nu} + \partial_{\nu}h_{\alpha\mu} - \partial_{\alpha}h_{\mu\nu}) + \bigcirc(h^2) \tag{1.5}$$

To first order in h, we must raise and lower all indices using the flat Minkowski tensor $\eta_{\mu\nu}$ and not the full tensor $g_{\mu\nu}$, that is,

$$h^{\mu}_{\nu} \equiv \eta^{\mu\varrho}h_{\varrho\nu} \qquad \partial^{\mu} \equiv \eta^{\mu\nu}\partial_{\nu} \qquad \Box^2 \equiv \eta^{\mu\nu}\partial_{\mu}\partial_{\nu} \tag{1.6}$$

With this understanding, equations (1.4) and (1.5) yield the first-order Ricci tensor:

$$R_{\mu\nu}^{(1)} \equiv \frac{1}{2}(\partial_\lambda\partial_\mu h_\nu^\lambda + \partial_\lambda\partial_\nu h_\mu^\lambda - \Box^2 h_{\mu\nu} - \partial_\mu\partial_\nu h_\lambda^\lambda) \tag{1.7}$$

and equation (1.2) becomes

$$\partial_\lambda\partial_\mu h_\nu^\lambda + \partial_\lambda\partial_\nu h_\mu^\lambda - \Box^2 h_{\mu\nu} - \partial_\mu\partial_\nu h_\lambda^\lambda = 16\pi G S_{\mu\nu} \tag{1.8}$$

where

$$S_{\mu\nu} = T_{\mu\nu} - \frac{1}{2}\eta_{\mu\nu}T \tag{1.9}$$

As in electromagnetism, we can exploit a gauge symmetry of the linearised theory to get rid of spurious degrees of freedom. In fact, if we change coordinates

$$x^\mu \to x'^\mu = x^\mu + \xi^\mu \tag{1.10}$$

$h_{\mu\nu}$ transforms to lowest order as

$$h_{\mu\nu}(x) \to h'_{\mu\nu}(x') = h_{\mu\nu}(x) - (\partial_\mu\xi_\nu + \partial_\nu\xi_\mu) \tag{1.11}$$

so, if $|\partial_\mu\xi_\nu|$ is of the same order of magnitude as $|h_{\mu\nu}|$ then the condition $|h_{\mu\nu}| \ll 1$ is preserved and therefore this is a diffeomorphism of the theory.

We can choose a coordinate transformation such that

$$g^{\mu\nu}\Gamma_{\mu\nu}^\lambda = 0 \tag{1.12}$$

This is called the harmonic, or De Donder gauge. Some texts call it also Lorenz gauge in analogy to electromagnetism. To first order it implies that:

$$\partial_\mu h_\nu^\mu = \frac{1}{2}\partial_\nu h_\mu^\mu \tag{1.13}$$

In this gauge, the linearised field equations (1.8) become

$$\Box^2 h_{\mu\nu} = -16\pi G S_{\mu\nu} \tag{1.14}$$

The solution to this equation is given by the retarded potential:

$$h_{\mu\nu}(\vec{x}, t) = 4G \int d^4x' \frac{S_{\mu\nu}(\vec{x}', t)}{|x - x'|} \tag{1.15}$$

In order to understand the physical content of equation (1.14) let us solve it in the region far away from the source, i.e.

$$\Box^2 h_{\mu\nu} = 0 \tag{1.16}$$

The most general solutions to equation (1.16) can be written as plane-wave solutions:

$$h_{\mu\nu}(\vec{x}, t) = e_{\mu\nu} \exp\left[i(\omega t - \vec{k} \cdot \vec{x})\right] + e_{\mu\nu}^* \exp\left[-i(\omega t - \vec{k} \cdot \vec{x})\right] \qquad (1.17)$$

This satisfies equation (1.16) if

$$k_\mu k^\mu = 0 \qquad (1.18)$$

and equation (1.13) if

$$k_\mu e_\nu^\mu = \frac{1}{2} k_\mu e_\nu^\nu \qquad (1.19)$$

The matrix $e_{\mu\nu}$ is obviously symmetric and is called the *polarisation tensor*.

This tensor is symmetric in its indices, so it has $n(n + 1)/2 = 4 \cdot 5/2 = 10$ independent components. The four relations (19) reduce this number to six, but of these six only two represent physically significant degrees of freedom. As we have seen, under a coordinate transformation $x^\mu \rightarrow x^\mu + \xi^\mu$ the metric $\eta_{\mu\nu} + h_{\mu\nu}$ transforms into a new metric $\eta_{\mu\nu} + h_{\mu\nu}'$, where $h_{\mu\nu}'$ is given by equation (1.11). If we choose:

$$\xi^\mu(x) = i\xi^\mu \exp\left(ik_\mu x^\mu\right) - i\xi^{\mu*}\exp\left(-ik_\mu x^\mu\right) \qquad (1.20)$$

then equation (1.11) gives

$$h_{\mu\nu}' = e_{\mu\nu}' \exp(ik_\lambda x^\lambda) + e_{\mu\nu}'^* \exp(-ik_\lambda x^\lambda) \qquad (1.21)$$

with

$$e_{\mu\nu}' = e_{\mu\nu} + k_\mu \xi_\nu + k_\nu \xi_\mu \qquad (1.22)$$

It is easy to see that $e_{\mu\nu}$ and $e'_{\mu\nu}$ represent the very same physical situation for arbitrary values of the four parameters ξ^μ, so, of the six independent components of $e_{\mu\nu}$, only $6 - 4 = 2$ represent physical degrees of freedom.

To illustrate this point and to make our notation uniform to that of many books and articles on GWs, let us consider, as an example, a wave travelling in the z-direction, i.e. with wave vector

$$k^1 = k^2 = 0 \qquad k^3 = k^0 = k > 0 \qquad (1.23)$$

In this case equation (1.19) gives

$$e_{31} + e_{01} = e_{32} + e_{02} = 0$$

$$e_{33} + e_{03} = -e_{03} - e_{00} = \frac{1}{2}(e_{11} + e_{22} + e_{33} - e_{00})$$

These equations allow us to express e_{i0} and e_{22} in terms of the other six $e_{\mu\nu}$:

$$e_{31} = -e_{01}; \qquad e_{02} = -e_{32}; \qquad e_{03} = \frac{1}{2}(e_{33} + e_{00}); \qquad e_{22} = -e_{11} \qquad (1.24)$$

When the coordinate system transforms according to equation (1.20) these six independent components transform according to equation (1.22):

$$e'_{11} = e_{11} \qquad e'_{12} = e_{12}$$

$$e'_{13} = e_{13} + k\xi_1 \qquad e'_{23} = e_{23} + k\xi_2$$

$$e'_{33} = e_{33} + 2k\xi_1 \qquad e'_{00} = e_{00} - 2k\xi_0 \qquad (1.25)$$

We see therefore that it is only e_{11} and e_{12} that have an absolute physical significance. We can arrange that all components of $e'_{\mu\nu}$ vanish except for e'_{11} and e'_{12} and $e'_{22} = -e'_{11}$ by performing a coordinate transformation with

$$\xi_1 = -\frac{e_{13}}{k}; \qquad \xi_2 = -\frac{e_{23}}{k}; \qquad \xi_3 = -\frac{e_{33}}{2k}; \qquad \xi_0 = \frac{e_{00}}{2k} \qquad (1.26)$$

The distinction between the different components of the polarisation tensor becomes clear if we study how $e_{\mu\nu}$ changes when we subject the coordinate system to a rotation about the z-axis, a Lorentz transformation of the form:

$$R_1^1 = \cos\theta; \qquad R_1^2 = \sin\theta$$

$$R_2^1 = -\sin\theta; \qquad R_2^2 = \cos\theta$$

$$R_3^3 = R_0^0 = 1 \qquad \text{other } R_\nu^\mu = 0 \qquad (1.27)$$

Since this transformation leaves k_μ invariant, the only effect is to transform $e_{\mu\nu}$ into

$$e'_{\mu\nu} = R_\mu^\alpha R_\nu^\beta e_{\alpha\beta} \qquad (1.28)$$

Using the relations (24) we find that:

$$e'_\pm = \exp(\pm 2i\theta)e_\pm \qquad (1.29)$$

$$f'_\pm = \exp(\pm i\theta)f_\pm \qquad (1.30)$$

$$e'_{33} = e_{33}; \qquad e'_{00} = e_{00} \qquad (1.31)$$

where

$$e_\pm \equiv e_{11} \mp ie_{12} = -e_{22} \mp ie_{12} \qquad (1.32)$$

$$f_\pm \equiv e_{31} \mp ie_{32} = -e_{01} \pm ie_{12} \qquad (1.33)$$

In general any plane wave ψ which is transformed by a rotation of any angle θ about the direction of propagation into

$$\psi' = \exp(ih\theta)\,\psi \tag{1.34}$$

is said to have **helicity** h. We have shown that a GW can be decomposed into parts $e_\pm$ with helicity ± 2, parts $f_\pm$ with helicity ± 1 and parts e_{00} and e_{33} with helicity zero. However, it is *only the parts with helicity ± 2 that have physical significance.*

We can write the tensor $h_{\mu\nu}$ as

$$h_{\mu\nu} = \begin{pmatrix} 0 & 0 & 0 & 0 \\ 0 & h_{11} & h_{12} & 0 \\ 0 & h_{12} & -h_{11} & 0 \\ 0 & 0 & 0 & 0 \end{pmatrix} \tag{1.35}$$

This is for a wave travelling along the z-direction. $h_{11} = -h_{22}$ is also referred to in literature as h_+, whereas $h_{12} = -h_{21}$ is called $h_\times$.

We can understand this argument in greater depth by exploring the analogy with electromagnetism. Maxwell's equations in empty space, together with the harmonic gauge condition, are:

$$\Box^2 A_\mu = 0; \qquad \partial_\mu A^\mu = 0 \tag{1.36}$$

A plane-wave solution is of the form:

$$A_\alpha = e_\alpha \exp(ik_\beta x^\beta) + e_\alpha^* \exp(-ik_\beta x^\beta) \tag{1.37}$$

where

$$k_\alpha k^\alpha = 0 \tag{1.38}$$

$$k_\alpha e^\alpha = 0 \tag{1.39}$$

In general e_α would have four independent components but the condition (1.39) reduces this number to three, just as equation (1.19) would reduce the number of independent $e_{\mu\nu}$ components from ten to six. Furthermore, without changing the physical fields $\vec{E}$ and $\vec{B}$ and without leaving the Lorenz gauge, we can change A_α by a gauge transformation:

$$A_\mu \rightarrow A'_\mu = A_\mu + \partial_\mu \Phi$$

$$\Phi(x) = i\xi \exp(ik_\lambda x^\lambda) - i\xi^* \exp(-ik_\lambda x^\lambda)$$

in analogy with equations (1.20) and (1.21). The new potential can be written:

$$A'_\alpha = e'_\alpha \exp(ik_\beta x^\beta) + e'^*_\alpha \exp(-ik_\beta x^\beta)$$

$$e'_\alpha = e_\alpha - \xi k_\alpha$$

The parameter ξ is arbitrary, so of the algebraically three independent components only $3 - 1 = 2$ are physically significant. To identify the two significant components of e_α we may consider a wave travelling in the z-direction, with k^α given by equation (1.23). The condition that $k_\mu e^\mu = 0$ allows us to determine e^0:

$$e_0 = -e_3$$

The preceding gauge transformation leaves e_1 and e_2 invariant, but changes e_3 into

$$e_3' = e_3 - \xi k$$

hence e_3' can be set equal to zero by choosing $\xi = e_3/k$ and so it is only e_1 and e_2 that carry physical significance. To work out the meaning of these two components we can subject the electromagnetic wave to the rotation defined by equation (1.27). The polarisation vector is then changed into:

$$e_\alpha' = R_\alpha^{\ \beta} e_\beta$$

and therefore

$$e_\pm' = \exp(\pm i\theta) e_\pm$$

$$e_3' = e_3$$

where

$$e_\pm \equiv e_1 \mp i e_2$$

Thus an electromagnetic wave can be decomposed into parts with helicity ± 1 and parts with helicity zero. However, the only physically significant ones are those with helicity ± 1, just as for GWs they are ± 2. This is what we mean when we say that electromagnetism and gravitation are carried by waves of spin 1 and spin 2, respectively.

It is a general fact that *waves that propagate with the speed of light have only two helicity states.* The hypotetical carrier of the gravitational interaction, the graviton, carries spin 2.

1.4 Background reading

My mathematical treatment of GR follows closely that of Steven Weinberg's *Gravitation and Cosmology* (Wiley 1972): although written 44 years ago, and with no exercises to solve, it remains a masterpiece, where the physical reasoning goes perfectly hand-in-hand with the mathematical formalism. Another excellent, and more modern book, about GR is *Introduction to General Relativity* by Lewis Ryder (Cambridge University Press 2009). As it concerns the SGWB, I am very grateful to Bruce Allen, whose lectures on the SGWB during a summer school back in 1999 in the beautiful location of Lake Como inspired me to devote my energies to the SGWB. Allen's lectures were published in the *Proceedings of the Les Houches on Astrophysical Sources of Gravitational Waves,* eds Jean-Alain Marck and Jean-Pierre Lesota (Cambridge University Press 1996).

References

[1] Abbott B P *et al* 2016 (LIGO Scientific Collaboration and Virgo Collaboration) Observation of gravitational waves from a binary black hole merger *Phys. Rev. Lett.* **116** 061102

[2] Abbott B P *et al* 2016 (LIGO Scientific Collaboration and Virgo Collaboration) GW151226: Observation of gravitational waves from a 22-solar-mass binary black hole coalescence *Phys. Rev. Lett.* **116** 241103

[3] Abbott B P *et al* 2017 (LIGO Scientific and Virgo Collaboration) GW170104: Observation of a 50-solar-mass binary black hole coalescence at redshift 0.2 *Phys. Rev. Lett.* **118** 221101

[4] Abbott B P *et al* 2017 (LIGO Scientific Collaboration and Virgo Collaboration) GW170814: A three-detector observation of gravitational waves from a binary black hole coalescence *Phys. Rev. Lett.* **119** 141101

[5] Abbott B P 2017 (LIGO Scientific Collaboration and Virgo Collaboration) *et al* GW170817: Observation of gravitational waves from a binary neutron star inspiral *Phys. Rev. Lett.* **119** 161101

Chapter 2

General theoretical considerations about the cosmological SGWB

We wrote in the previous sections that the importance of the SGWB as a source of gravitational radiation lies in the fact that, with the exception of a component that stems from the random superposition of many weak binary systems, henceforth called '(astrophysical) foreground', the SGWB is the result of processes that took place in the very early Universe, when the Universe was less than a picosecond old. We lack a quantum theory of gravity, yet we can try to calculate heuristically when such a component was produced. For any given particle species, the decoupling time t_{dec} is defined as the time when its interaction rate Γ equals the expansion rate of the Universe, as measured by the Hubble parameter $H(t)$. In fact it can be shown, for all particles for which we have a sound quantum description, i.e. Standard Model particles, that the number of interactions the particle species has after t_{dec} is less than one. Since the particle species has no interactions with the rest of the Universe after the decoupling time, its spectrum retains memory of the Universe as it was at t_{dec}. The only change in the spectrum is the redshift of the magnitude of the three-momentum due to the expansion of the Universe.

At a given temperature T the interaction rate for particles that interact only gravitationally can be written, on purely dimensional reasoning, as

$$\Gamma \sim G_N^2 T^5 = \frac{T^5}{M_{Pl}^4} \qquad (2.1)$$

where G_N is Newton's constant and $M_{Pl} = 1/\sqrt{G_N} = 1.22 \times 10^{19}$ GeV is the Planck mass. In the radiation dominated era, $H \sim T^2/M_{Pl}$ which means that the decoupling temperature T_{dec} is given by

$$T_{dec} \sim M_{Pl} \qquad (2.2)$$

doi:10.1088/978-1-6817-4082-9ch2

Gravitons decouple at the Planck scale. Of course, since we have no quantum theory of gravity, this argument can be seen as just an approximation. It is, however, very similar to the calculations performed by Fermi back in the 1930s for the cross section of neutrinos, estimates that proved to be an excellent approximation to the real values, even after the right quantum theory of the weak interaction was formulated. Just like the CMBR mentioned earlier, the SGWB is a randomly polarised relic of the early Universe. The difference however is that we step back from $t_{\text{dec}}^{\text{el}} = 4 \times 10^5$ years to the time that characterises the Planck scale, $t_{\text{Pl}} = 5 \times 10^{-44}$ s. We underline that although our reasoning rests on shaky ground because of quantum gravity, the physics at such high energies cannot be accessed experimentally in any other way.

String theory has long been seen as a viable theory of quantum gravity: according to it, the Planck mass is not a fundamental quantity, but results from a more fundamental parameter, the so called string mass, after dimensional reduction. The string mass is somewhat smaller than the Planck mass, it lies presumably in the $10^{17}-10^{18}$ GeV region, therefore, the corresponding decoupling time would be one or two orders of magnitude larger than t_{Pl}, we are still talking, however, about fractions of a second, well below the picosecond scale.

From an experimental point of view, the frequency range covered by ground-based detectors such as LIGO and Virgo is 4 Hz–10 kHz. 100 Hz is the frequency where LIGO detectors are most sensitive: taking into account the redshift due to the Universe's expansion, a gravitational radiation that lies today in the 100 Hz frequency region had to be produced when the temperature of the Universe was $T_* \sim 6 \times 10^8$ GeV, which corresponds to a time $t_* \sim 7 \times 10^{-25}$ s. This is shortly after inflation, so we would have a picture of the Universe entering the radiation dominated era.

These estimates, we repeat once more, are sound from a physical point of view, but are not very accurate mathematically since there is no framework in which to perform exact calculations. However, gravitons below the Planck scale interacted too weakly to thermalise, so there is no reason to assume the usual f^2 dependence—typical of thermal spectra—for their spectrum. The form of their spectrum is model dependent; flat or almost flat spectra over a large range of frequencies are favoured, so even if our naive arguments should prove wrong, we can still hope to detect the long low-frequency tails of these spectra with our ground-based detectors.

It is useful to describe the spectrum of the SGWB by specifying how energy is distributed in frequency. We will not delve too deply into the vexed issue of how energy is defined in GR [1]. In the weak-field approximation it is possible to give a meaningful, local definition of energy, which however cannot be carried over to the full theory, and we will label this quantity as ρ_{GW}. The intensity of the stochastic background is thus given by the dimensionless quantity

$$\Omega_{\text{GW}}(f) \equiv \frac{1}{\rho_{\text{c}}} \frac{\text{d}\rho_{\text{GW}}}{\text{d}\ln f} \qquad (2.3)$$

where f is the frequency and ρ_c is the present value of energy density required to close the Universe. In terms of the Hubble constant, H_0 and keeping natural units, i.e. units for which $c = 1$, ρ_c is given by

$$\rho_c \equiv \frac{3H_0^2}{8\pi G_N} \tag{2.4}$$

G_N being Newton's constant.

There has been some confusion in the literature about $\Omega_{GW}(f)$. Some authors assume $\Omega_{GW}(f)$ is independent of frequency. Although this is indeed the case in some cosmological models, it is not true for all of them. The main point is that any GW spectrum can be described by an appropriate $\Omega_{GW}(f)$.

We do not have an accurate value for H_0, also a matter of consistent controversy in literature. Because of the difficulties in measuring galactic distances, estimates of H_0 have changed by over an order of magnitude since Hubble's original work. Recent measurements [1, 2] give values consistent with

$$H_0 = (70 \pm 10) \text{ km s}^{-1} \text{ Mpc}^{-1} \tag{2.5}$$

Since H_0 is omnipresent in cosmological formulas, it is very useful for numerical results to shift the uncertainty on to a new dimensionless parameter h_{100}, which lies almost certainly in the $1/2 < h_{100} < 1$ range, defined as $h_{100} = H_0/(100 \text{ km s}^{-1} \text{ Mpc}^{-1})$:

$$H_0 = 100h_{100} \text{ km s}^{-1} \text{ Mpc}^{-1} \tag{2.6}$$

While h is widely used in literature and in most articles about the SGWB we prefer to use a slightly different parameter, namely

$$h_{70} = H_0/70(\text{ km s}^{-1} \text{ Mpc}^{-1}) \tag{2.7}$$

from which we have

$$H_0 = 70h_{70} \text{ km s}^{-1} \text{ Mpc}^{-1} \qquad h_{70} = 1.0 \pm 0.15 \tag{2.8}$$

which will give numerical factors in line with present measurements of H_0 simply by assuming $h_{70} \sim 1$. Since Ω_{GW} depends on H_0 and therefore on its uncertainty, we will characterise the SGWB through the quantity $h_{70}^2\Omega_{GW}$ which is independent of the uncertainty on H_0. Our choice of h_{70}, which is roughly one, rather than simply h, allows us to 'forget' about h_{70} and write only Ω_{GW}, but we must not forget that this is shorthand for the real quantity $h_{70}^2\Omega_{GW}$.

It is useful to relate $\Omega_{GW}(f)$ to a measurable quantity often used by experimentalists, namely the dimensionless strain measured by interferometric detectors $h = \Delta L/L$. The two quantities are related by the following formula:

$$h = 3 \times 10^{-20}h_{70}\sqrt{\Omega_{GW}(f)}\frac{100 \text{ Hz}}{f} \tag{2.9}$$

It is easy to see that if $\Omega_{GW}(f) = 10^{-8}$ in a bandwidth 50 Hz $< f <$ 100 Hz, then the ideal strain produced in a detector is $h \approx 10^{-24}$.

$h_{70}^2 \Omega_{GW}(f)$ contains all the information about the SGWB if we make further assumptions, namely that the SGWB is isotropic, stationary, unpolarised and Gaussian: assumptions we are going to discuss now.

It is experimentally established that the CMBR is highly isotropic [1]. CMBR temperature fluctuations across the sky are at the level $\Delta T/T \sim 10^{-5}$ and arise because of the non uniform distribution of matter at the surface of last scattering. In fact this isotropy is very surprising because the size of the causal horizon at recombination ($z = 1100$, where z is the redshift) is seen today under the angle of about 1°. It is therefore very puzzling that we see the same microwave background temperature on patches separated by more than 1°, which have never been in causal contact before the microwave photons were emitted. This mystery goes under the name of 'horizon problem' in the Standard Model of Cosmology and is the main reason that has led to the formulation of the inflationary model. There is also a spurious dipole effect $\Delta T/T \sim 10^{-3}$ due to the motion of our Solar System with respect to the rest frame of the Universe. It is therefore reasonable to expect that a stochastic background of cosmological origin, generated at earlier times, i.e. at higher redshift than the CMBR, should be also isotropic at least in a first approximation. Ot course, after the SGWB has been detected, it will be extremely interesting to investigate its anisotropies. We will certainly find the dipole term, dominated by the Earth motion in the rest frame of the CMBR, and the higher multipoles could give extremely interesting information about the very early Universe.

We might have to give up the isotropy assumption when we study a background of astrophysical origin, since galaxies are not isotropically distributed. Such a foreground, created by galactic white dwarf binaries, would be more intense when we look in the direction of the galactic plane, exactly in the same way as the electromagnetic background due to galactic sources gives the Milky Way its typical appearance.

The assumption that the background is stationary is almost certainly justified. Technically this means that the n-point correlation functions of the GW fields depend only on the differences between the times and not on the choice of the time origin. Because the age of the Universe is 20 orders of magnitude larger than the characteristic period of the waves LIGO, Virgo and other ground-based facilities can detect and 9 orders of magnitude larger than the longest realistic observation times, it seems very unlikely that the SGWB can have properties that vary over either of these time scales. The instrumental noise however, which enters the picture when we discuss about detection strategies, is not stationary. We will talk about the issue later.

We make the assumption that the background is Gaussian, which means that the joint density function is a multivariate normal distribution. Gaussianity is rooted in the central limit theorem which states that the sum of a large number of independent events produces a Gaussian stochastic process regardless of the probability distributions of the individual events. The number of separate horizon volumes which give rise to a stochastic background at $t = 10^{-22}$ s is $N_{\text{horizon}} = z^2 \sim 10^{39}$. Whether or not this assumption would hold true also for a foreground of astrophysical origin is debated; we remind the reader that already with $n = 10$ variables a Poissonian probability distribution is well approximated by a Gaussian.

Finally we assume that the SGWB is unpolarised, which is natural both in a cosmological context and also if it is the result of the superposition of many different astrophysical sources.

2.1 The SGWB and its detectability

To detect an SGWB component, be it of cosmological or astrophysical origin, the optimal strategy consists in correlating the output of two detectors, since the expected signal is not only far too low to stand above the noise threshold but also looks exactly like an additional source of noise. Let us discuss briefly why. Experimentalists can carefully model the noise of a detector and give a plausible value of the *noise spectral density* $S_n(f)$. When a detector is turned on, its output is always

$$s(t) = n(t) + h(t) \qquad (2.10)$$

with $n(t)$ the instrumental noise and $h(t)$ the response of the detector to a true GW signal. We measure $\langle s^2(t) \rangle$ and if it turns out to be much larger than expected—in a sense we will clarify shortly—the crucial problem becomes how to tell between a GW background and an unaccounted, or larger than expected, source of noise. Penzias and Wilson met with the very same problem: they found excess noise in their antenna for satellite communications and had to work hard for more than a year to discard all possible explanations in terms of an unknown noise source. Their short paper—the second shortest to be awarded the Nobel Prize—had the very modest title 'A measurement of excess antenna temperature at 4080 Mc s^{-1}' [3] and the article's (famous) last words were: 'From a combination of the above, we compute the remaining unaccounted-for antenna temperature to be 3.5 ± 1.0 K at 4080 Mc s^{-1}'.

To deal with this problem, the most clever strategy is to set a very high threshold on the signal-to-noise-ratio (SNR), let's say $S/N = 5$. At the same time we could look for possible spatial anisotropies in the SGWB component of astrophysical origin, due to the sidereal time modulation caused by the motion in the Solar System of the detector with respect to the cosmological rest frame. And we could look for a specific frequency dependency on the signal, as predicted by some cosmological or high-energy physics theories. But without those handles, we must set a high SNR threshold. Without showing all the details of the calculation—the interested reader can look it up in [4]—we quote the result that, for a given noise spectral density $S_n(f)$ the minimum detectable signal is

$$[\Omega_{\text{GW}}(f)]_{\text{min}} = \frac{4\pi^2}{3H_0^2} f^3 S_n(f) \frac{(S/N)^2}{F} \qquad (2.11)$$

where F is the so-called *angular efficiency factor*, a number which depends on the geometry of the detectors and whose value for ground-based interferometers is roughly 2/5.

The essential feature in equation (2.11) is the f^3 factor. It tells us that if we are able to reach a given level in $S_n(f)$ at low frequencies it will be possible to reach a much better sensitivity in $\Omega_{GW}(f)$ compared to what could be obtained at higher frequencies with the same $S_n(f)$. It is much easier to reach a small level for $[\Omega_{GW}(f)]_{min}$ at low frequencies than at high frequencies. There is a large variety of possible mechanisms which can produce SGWB components everywhere from $f = 10^{-18}$ Hz up to $f = 10^9$ Hz. Their detection is therefore easier at low frequencies and this is the main reason why future space-based missions such as eLISA (enhanced Laser Interferometric Space Antenna) are so important, since they will probe the low-frequency region (10^{-4} Hz to 1 Hz).

So, apart from very strong and very unlikely signals, if we want to detect the SGWB in the 4 Hz to 10 kHz region, where ground-based interferometric detectors are most sensitive, we need at least two detectors. We are now going to describe the optimal data analysis strategy.

2.1.1 Expectation value

Much of what follows here is taken from the famous article by Bruce Allen and Joseph Romano [5] which marked a milestone in the literature for the SGWB and expanded the data analysis section of the above mentioned lectures by Bruce Allen.

Our starting point is the plane-wave expansion of the gravitational metric perturbation $h_{\mu\nu}$ in a transerse, traceless gauge (hence the indices will be labeled by latin letters instead of greek):

$$h_{ij}(t, \vec{x}) = \sum_A \int_{-\infty}^{+\infty} df \int_{S^2} d\hat{\Omega} h_A(f, \hat{\Omega}) \ \exp[2\pi i f(t - \hat{\Omega} \cdot \vec{x})] e_{ij}^A(\hat{\Omega}) \qquad (2.12)$$

where $\hat{\Omega}$ is a unit vector that specifies a direction on the two-sphere, with wave vector $k \equiv 2\pi f \hat{\Omega}$ (we remind the reader that, unless otherwise specified, we have set $c = 1$). $A = +,\times$ denotes the polarisation states and the corresponding polarisation tensors $e_{ij}^A(\hat{\Omega})$ are given by:

$$e_{ij}^+(\hat{\Omega}) = \hat{m}_i \hat{m}_j - \hat{n}_i \hat{n}_j \qquad (2.13)$$

$$e_{ij}^\times(\hat{\Omega}) = \hat{m}_i \hat{n}_j + \hat{n}_i \hat{m}_j \qquad (2.14)$$

where

$$\hat{\Omega} = \cos\phi \sin\theta \hat{x} + \sin\phi \sin\theta \hat{y} + \cos\theta \hat{z} \qquad (2.15)$$

$$\hat{m} = \sin\phi \hat{x} - \cos\phi \hat{y} \qquad (2.16)$$

$$\hat{n} = \cos\phi \cos\theta \hat{x} + \sin\phi \cos\theta \hat{y} - \sin\theta \hat{z} \qquad (2.17)$$

and (θ, ϕ) are the standard polar and azimuthal angles on the two-sphere. The Fourier amplitudes $h_A(f, \hat{\Omega})$ are arbitrary complex functions bound to satisfy $h_A(f, \hat{\Omega}) = h_A^*(f, \hat{\Omega})$, where * denotes as usual complex conjugation. This bound follows naturally from the reality of $h_{ij}(t, \vec{x})$.

The above stated assumptions that the stochastic background be isotropic, stationary, Gaussian and unpolarised imply that the expectation values of the Fourier amplitudes $h_A(f, \hat{\Omega})$ satisfy

$$\langle h_A^*(f, \hat{\Omega}) h_{A'}(f', \hat{\Omega}') \rangle = \delta^2(\hat{\Omega}, \hat{\Omega}') \delta_{A,A'} \delta(f - f') H(f) \tag{2.18}$$

where $\delta^2(\hat{\Omega}, \hat{\Omega}') = \delta(\phi - \phi') \delta(\cos\theta - \cos\theta')$ is the usual covariant Dirac delta distribution on the two-sphere and $H(f)$ is a real, non-negative function satisfying $H(f) = H(-f)$. If we further assume that the background has zero mean then

$$\langle h_A(f, \hat{\Omega}) \rangle = 0 \tag{2.19}$$

Since we are assuming the background to be Gaussian, equations (2.18) and (2.19) completely specify the background, all higher order moments being zero.

$H(f)$ introduced in equation (2.18) is related to the gravitational spectrum $\Omega_{GW}(f)$. The quantity ρ_{GW} we mentioned in the previous section can be defined as [4]:

$$\rho_{GW} = \frac{1}{32\pi G} \left\langle \dot{h}_{ij}(t, \vec{x}) \dot{h}^{ij}(t, \vec{x}) \right\rangle \tag{2.20}$$

By differentiating the plane-wave expansion (2.12) and taking the contractions, equations (2.19), (2.18) and (2.3) give:

$$\rho_{GW} = \frac{4\pi^2}{G} \int_0^\infty \mathrm{d}f f^2 H(f) \tag{2.21}$$

from which it immediately follows:

$$H(f) = \frac{3H_0^2}{32\pi^3} |f|^{-3} \Omega_{GW}(f) \tag{2.22}$$

and thus:

$$\left\langle h_A^*(f, \hat{\Omega}) h_{A'}(f', \hat{\Omega}') \right\rangle = \frac{3H_0^2}{32\pi^3} \delta^2(\hat{\Omega}, \hat{\Omega}') \, \delta_{A,A'} \delta(f - f') |f|^{-3} \Omega_{GW}(f) \tag{2.23}$$

which is the desired result.

2.1.2 Two detector theory: coincident and coaligned detectors

The most clever strategy to detect a random SGWB signal is to correlate the output of two detectors. We start by looking at the simplest case: two coincident and coaligned detectors. This is of course an ideal case, real-life complications will be gradually introduced in due course. The two outputs are given by:

$$s_1(t) = h_1(t) + n_1(t) \tag{2.24}$$

$$s_2(t) = h_2(t) + n_2(t) \tag{2.25}$$

$h_1(t)$ and $h_2(t)$ are the gravitational strains due to the stochastic background in the two detectors. Since we are assuming that the two detectors are coincident and coaligned, these strains must be the same, i.e.

$$h_1(t) = h_2(t) = h \tag{2.26}$$

The noise terms intrinsic to the individual detectors, $n_1(t)$ and $n_2(t)$, on the other hand, need not be the same. We will assume that the noise terms are stationary, Gaussian, uncorrelated and much larger in magnitude than the signal. These assumptions are bold, at least to a certain degree: we certainly do not expect noise to be stationary for all the lifetime of a detector ($\sim$10 years), but we can always find a given time interval—often very short—for which the noise is stationary and perform separate analysis for these shorter intervals. Instrumental noise comes from many different sources, so the Gaussianity is again given by the central limit theorem. That the noise should be much larger in magnitude than the signal is the starting assumption of our analysis and, given all the theoretical predictions about possible SGWB components, is certainly justified. We cannot expect however two coincident and coaligned detectors to have uncorrelated noise. This assumption makes our analysis much simpler, but we will relax this condition later on.

Given the two detector outputs equation (2.24) and equation (2.25) we can form a 'signal product' by multiplying them and integrating over time:

$$S \equiv \int_{-T/2}^{T/2} dt s_1(t) s_2(t) \tag{2.27}$$

where T is a given observation time. Since $s_1(t)$ and $s_2(t)$ are both random variables, so too is S. We define its expectation value μ as

$$\mu \equiv \langle S \rangle \tag{2.28}$$

and its variance is given by:

$$\sigma^2 \equiv \langle S^2 \rangle - \langle S \rangle^2 \tag{2.29}$$

with μ and σ^2 obviously related to the mean and variance of h, n_1 and n_2. The goal is to evaluate μ and σ^2 and then construct the SNR as:

$$\text{SNR} \equiv \frac{\mu}{\sigma} \tag{2.30}$$

This quantity will be of crucial importance in deciding whether to accept or reject a signal as a gravitational-wave signal. Let's start by calculating μ:

$$\mu = \langle S \rangle = \int_{-T/2}^{+T/2} dt \langle s_1(t)s_2(t) \rangle$$
$$= \int_{-T/2}^{+T/2} dt \langle h^2(t) + h_1(t)n_2(t) + n_1(t)h_2(t) + n_1(t)n_2(t) \rangle \tag{2.31}$$
$$= \int_{-T/2}^{+T/2} dt \langle h^2(t) \rangle$$
$$= T \langle h^2(t) \rangle = T \sigma_h^2$$

where σ_h^2 is the time independent variance of $h(t)$. In the third row, we have made use of the fact that signal and noise are obviously uncorrelated as are by assumption the two noise terms:

$$\langle h_1(t)n_2(t) \rangle = \langle n_1(t)h_2(t) \rangle = \langle n_1(t)n_2(t) \rangle = 0 \tag{2.32}$$

To express σ_h^2 in terms of $\Omega_{GW}(f)$ we will make use of the plane-wave expansion (2.12). Since

$$h(t) \equiv h_{ij}(t, \vec{x}_0) \frac{1}{2} (\hat{X}^i \hat{X}^j - \hat{Y}^i \hat{Y}^j) \tag{2.33}$$

where $\vec{x}_0$ is the position of the central station of the two coincident detectors and $\hat{X}^i$ and $\hat{Y}^j$ are unit vectors pointing to the direction of the arms, it follows that

$$\sigma_h^2 = \sum_A \sum_{A'} \int_{S^2} d\hat{\Omega} \int_{S^2} d\hat{\Omega}' \int_{-\infty}^{+\infty} df \int_{-\infty}^{+\infty} df' \left\langle h_A^*(f, \hat{\Omega}) h_{A'}'(f', \hat{\Omega}') \right\rangle$$
$$\times \exp\left[-i2\pi f(t - \hat{\Omega} \cdot \vec{x}_0)\right] \exp\left[i2\pi f'(t - \hat{\Omega}' \cdot \vec{x}_0)\right] \tag{2.34}$$
$$\times F^A(\hat{\Omega}) FA'(\hat{\Omega}')$$

where

$$F^A(\hat{\Omega}) \equiv e_{ij}^A(\hat{\Omega}) \frac{1}{2} (\hat{X}^i \hat{X}^j - \hat{Y}^i \hat{Y}^j) \tag{2.35}$$

is the response of the individual detector to a zero frequency, unit amplitude, $A = +, \times$ GW.

Using equation (2.18), equation (2.27) becomes:

$$\sigma_h^2 = \frac{3H_0^2}{32\pi^3} \sum_A \int_{-\infty}^{+\infty} df \, |f|^{-3} \Omega_{GW}(|f|) \int_{S^2} d\hat{\Omega} F^A(\hat{\Omega}) F^A(\hat{\Omega})$$
$$= \frac{3H_0^2}{20\pi^2} \int_{-\infty}^{+\infty} df \, |f|^{-3} \Omega_{GW}(|f|) \tag{2.36}$$

where we have made use of

$$\sum_A \int_{S^2} d\hat{\Omega} F^A(\hat{\Omega}) F^A(\hat{\Omega}) = \frac{8\pi}{5} \tag{2.37}$$

For coincident, coaligned detectors we thus obtain

$$\mu = T\sigma_h^2 = \frac{3H_0^2}{20\pi^2} T \int_{-\infty}^{+\infty} df \, |f|^{-3} \Omega_{\text{GW}}(|f|) \qquad (2.38)$$

This is the first result. Let us now calculate the variance $\sigma^2 = \langle S^2 \rangle - \langle S \rangle^2$:

$$
\begin{aligned}
\sigma^2 &= \langle S^2 \rangle - \langle S \rangle^2 \\
&= \int_{-T/2}^{+T/2} dt \int_{-T/2}^{+T/2} dt' \langle s_1(t)s_2(t)s_1(t')s_2(t') \rangle - \left[\int_{-T/2}^{T/2} dt \langle s(t) \rangle \right]^2 \\
&= \int_{-T/2}^{+T/2} dt \int_{-T/2}^{+T/2} dt' \langle n_1(t)n_2(t)n_1(t')n_2(t') \rangle \\
&= \int_{-T/2}^{+T/2} dt \int_{-T/2}^{+T/2} dt' \langle n_1(t)n_1(t') \rangle \langle n_2(t)n_2(t') \rangle
\end{aligned}
\qquad (2.39)
$$

In the third line, we used the fact that signal and noise are uncorrelated and that, as calculated previously, the variance of $h(t)$, $\langle h^2(t) \rangle$ is independent of time. In the last line, we used the assumption of statistical independence of the noise in the two detectors.

By definition of instrumental noise we have:

$$\langle n_i(t)n_i(t') \rangle \equiv \frac{1}{2} \int_{-\infty}^{+\infty} df \, \exp\left[i2\pi f(t - t')\right] P_i(|f|) \qquad (2.40)$$

where $P_i(|f|)$ is the one-sided *noise power spectrum* of the ith-detector. It is a real, positive-definite function, which satisfies:

$$\langle n_i^2(t) \rangle = \int_0^{+\infty} df P_i(f) \qquad (2.41)$$

which states that the *total noise power* is the integral of $P_i(f)$ over the positive frequencies, hence the factor $1/2$ in equation (2.40).

Equation (2.40) can also be written in the frequency domain:

$$\langle \tilde{n}_i^*(f)\tilde{n}_i(f') \rangle = \frac{1}{2}\delta(f - f')P_i(|f|) \qquad (2.42)$$

Inserting equation (2.40) in equation (2.39) gives:

$$
\begin{aligned}
\sigma^2 &= \frac{1}{4} \int_{-T/2}^{+T/2} dt \int_{-T/2}^{+T/2} dt' \int_{-\infty}^{+\infty} df \int_{-\infty}^{+\infty} df' \\
&\quad \times \exp\left[i2\pi f(t - t')\right] \exp\left[-i2\pi f'(t - t')\right] P_1(|f|) P_2(|f'|)
\end{aligned}
\qquad (2.43)
$$

Integrating this expression over t and t' we find

$$\sigma^2 = \frac{1}{4} \int_{-\infty}^{+\infty} df \int_{-\infty}^{+\infty} df' \delta_T^2(f - f')P_1(|f|)P_2(|f|) \qquad (2.44)$$

where

$$\delta_T(f) \equiv \int_{-T/2}^{+T/2} dt \, \exp(-i2\pi ft) \tag{2.45}$$

is a finite-time approximation to the Dirac delta function $\delta(f)$. In practice, we can replace one $\delta_T(f)$ function with a Dirac delta function and calculate the other in $f' = f$. Doing so yields:

$$\sigma^2 = \frac{1}{4} T \int_{-\infty}^{+\infty} df P_1(|f|) P_2(|f|) \tag{2.46}$$

We are now fully equipped to calculate the SNR

$$\text{SNR} = \frac{\mu}{\sigma} = \frac{3H_0^2}{10\pi^2} \sqrt{T} \frac{\int_{-\infty}^{+\infty} df \, |f|^{-3} \Omega_{\text{GW}}(|f|)}{\left[\int_{-\infty}^{+\infty} df P_1(|f|) P_2(|f|) \right]^{1/2}} \tag{2.47}$$

Equation (2.47) is the most important result in SGWB data analysis. It says that, no matter how small the signal $\Omega_{\text{GW}}(f)$ or how noisy the detectors, by integrating over a sufficiently long time T we can exceed any prescribed value of the SNR. Although this result has been obtained in the ideal case of two coincident and coaligned detectors with uncorrelated noise, it stands also in the more general case.

2.1.3 The overlap reduction function

When we consider the correlation between the outputs of two GW interferometers we must take into account that in general two detectors are not necessarily coincident and coaligned. There will be a reduction in sensitivity due to: i) the separation time delay between the two detectors and ii) the non parallel alignement of the detector arms

The two effects mentioned above imply that $h_1(t)$ and $h_2(t)$ are no longer the same, the overlap between the gravitational strains in the two detectors is only partial. To take it into account, we introduce the *overlap reduction function* (ORF), a dimensionless function of the frequency defined as:

$$\gamma(f) \equiv \frac{5}{8\pi} \sum_A \int_{S^2} d\hat{\Omega} \, \exp(i2\pi f \hat{\Omega} \cdot \Delta \vec{x}) F_1^A(\hat{\Omega}) F_2^A(\hat{\Omega}) \tag{2.48}$$

where $\Delta \vec{x}$ is the spatial separation between the central stations of the two sites and

$$F_i^A(\hat{\Omega}) \equiv e_{ab}^A(\hat{\Omega}) d_i^{ab} = e_{ab}^A(\hat{\Omega}) \frac{1}{2} \left(\hat{X}_i^a \hat{X}_i^b - \hat{Y}_i^a \hat{Y}_i^b \right) \tag{2.49}$$

is the response of the ith detector ($i = 1, 2$) to the $A = +, \times$ polarisation mode, that we had defined previously also in the case of coincident and coaligned detectors by equation (2.35). The ORF $\gamma(f)$ equals unity for all frequencies for coincident and coaligned detectors. It decreases when the two detectors are shifted apart or when

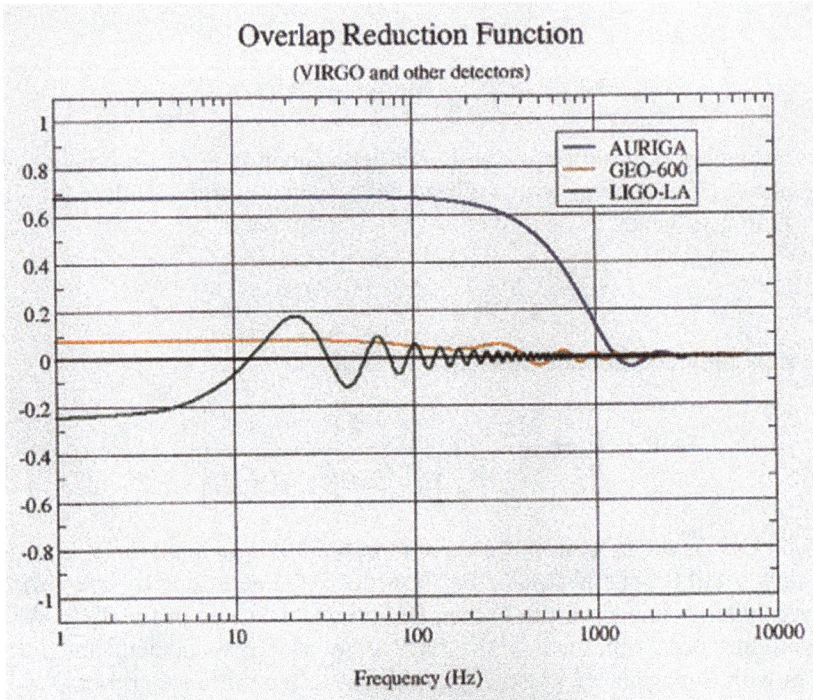

Figure 2.1. Virgo's overlap reduction function with some of the other detectors.

they are rotated out of coalignement and so respond differently to the polarisation states. In figure 2.1, courtesy of Michele Maggiore, we show, as an example, Virgo's ORF with some of the other detectors (AURIGA is a bar detector no longer in operation, but it was the closest detector to Virgo geographically speaking). I have chosen Virgo because I am Italian! Figure 2.2 shows the overlap reduction function for LIGO-LA. Allegro, a bar detector, is no longer in operation, but I have decided to leave it in the graph because LIGO-LA and Allegro were a pair of detectors very close geographically speaking and so their overlap reduction function is very instructive to see.

2.1.4 Optimal filtering

All detections of GWs are based on some form of *matched filtering*, we apply to the detector output a filter that enhances the signal with respect to the noise. The main requirement is that we must know the waveform of the signal for which we are looking. The techniques laid out in the previous section allow us to derive a rigorous strategy.

The most general form of equation (2.27) can be written as:

$$S = \int_{-T/2}^{+T/2} dt \int_{-T/2}^{+T/2} dt' s_1(t) s_2(t') Q(t, t') \tag{2.50}$$

where as before (see equations (2.24) and (2.25))

Overlap Reduction Function
(LIGO-LA and other detectors)

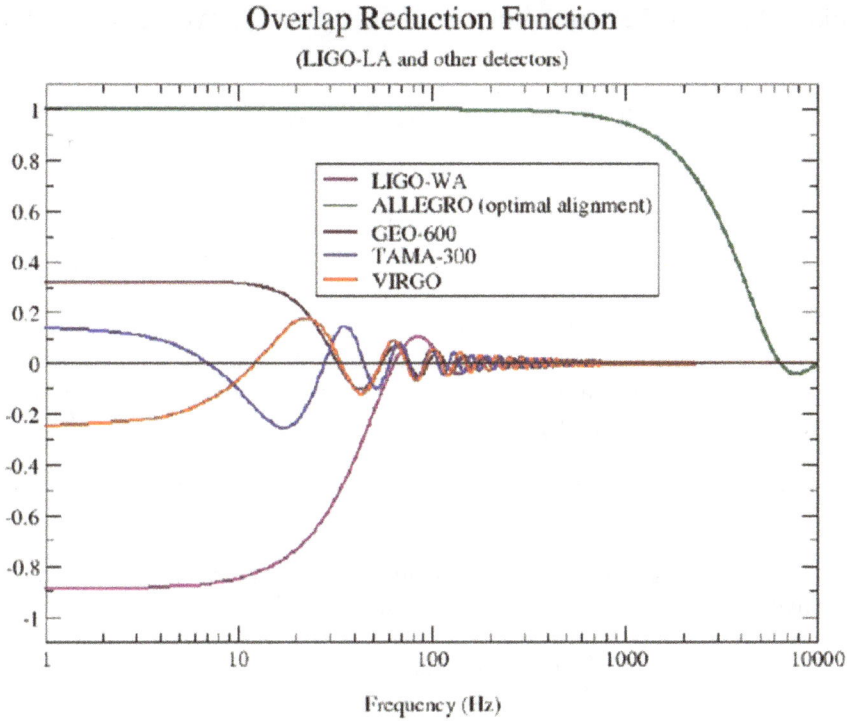

Figure 2.2. Overlap reduction function of the LIGO LA detector (courtesy of the LIGO Virgo Collaboration).

$$s_1(t) = h_1(t) + n_1(t)$$
$$s_2(t) = h_2(t) + n_2(t) \quad (2.51)$$

but this time $Q(t, t')$ is now a filter function to be determined and not simply $\delta(t - t')$ as in equation (2.27). Given the assumed stationarity of both the signal and the noise—assumptions we discussed in a previous section—the best choice for the filter function can depend only on the time difference $\Delta t = t - t'$:

$$Q(t, t') = Q(t - t') \quad (2.52)$$

The goal is to find the *optimal filter function* $Q(t - t')$. The filter function will depend not only upon the SGWB spectrum and upon the noise power spectrum, but also on the locations and orientations of the detectors. It falls off rapidly to zero for time delays $\Delta t = t - t'$ much larger than the light travel time between the sites, d. Since a typical observation time T will also be much larger than d, we can change the integration limits in one of the integrals of equation (2.50). That will not change the physics and will make the maths more tractable:

$$S = \int_{-T/2}^{+T/2} dt \int_{-\infty}^{+\infty} dt' s_1(t) s_2(t') Q(t - t') \quad (2.53)$$

Equation (2.53) can be written in the Fourier domain. We define the Fourier transform as:

$$\tilde{g}(f) = \int_{-\infty}^{+\infty} dt\, g(t) \exp(-i2\pi ft) \tag{2.54}$$

and obtain thus

$$S = \int_{-\infty}^{+\infty} df \int_{-\infty}^{+\infty} df'\, \delta_T(f - f')\tilde{s}_1^*(f)\tilde{s}_2(f')\tilde{Q}(f) \tag{2.55}$$

Since $Q(t - t')$ is a real function, $\tilde{Q}(-f) = \tilde{Q}^*(f)$.

The natural question is: what is the optimal filter function with respect to the quantity we want to maximise, i.e. the SNR? We will calculate μ following the same reasoning that led us to equation (2.38).

$$\mu = \langle S \rangle = \int_{-\infty}^{+\infty} df \int_{-\infty}^{+\infty} df'\, \delta_T(f - f') \left\langle \tilde{h}_1^*(f)\tilde{h}_2(f') \right\rangle \tilde{Q}(f') \tag{2.56}$$

To calculate the expectation value $\langle \tilde{h}_1^*(f)\tilde{h}_2(f') \rangle$ we make use once more of the plane-wave expansion (12). Since ($i = 1, 2$ labels the detectors):

$$\tilde{h}_i(f) = \sum_A \int_{S^2} d\hat{\Omega}\, h_A(f, \hat{\Omega}) \exp(-i2\pi f \hat{\Omega} \cdot \vec{x}_i) F_i^A(\hat{\Omega}) \tag{2.57}$$

we find:

$$
\begin{aligned}
\left\langle \tilde{h}_1^*(f)\tilde{h}_2(f') \right\rangle &= \sum_A \sum_{A'} \int_{S^2} d\hat{\Omega} \int_{S^2} d\hat{\Omega}' \langle h_A^*(f, \hat{\Omega}) h_{A'}(f', \hat{\Omega}') \rangle \\
&\quad \times \exp(i2\pi f \hat{\Omega} \cdot \vec{x}_1) \exp(-i2\pi f'\hat{\Omega}' \cdot \vec{x}_2) F_1^A(\hat{\Omega}) F_2^{A'}(\hat{\Omega}') \\
&= \frac{3H_0^2}{32\pi^3} \delta(f - f')\, |f|^{-3}\, \Omega_{\mathrm{GW}}(|f|) \\
&\quad \sum_A \int_{S^2} d\hat{\Omega} \exp(i2\pi f \hat{\Omega} \cdot \Delta\vec{x}) F_1^A(\hat{\Omega}) F_2^A(\hat{\Omega}) \\
&= \frac{3H_0^2}{20\pi^2} \delta(f - f')\, |f|^{-3}\, \Omega_{\mathrm{GW}}(|f|)\gamma(|f|)
\end{aligned}
\tag{2.58}
$$

Substituting equation (2.58) into equation (2.56) gives:

$$\mu = \frac{3H_0^2}{20\pi^2} T \int_{-\infty}^{+\infty} df\, |f|^{-3}\, \Omega_{\mathrm{GW}}(|f|)\gamma(|f|)\tilde{Q}(f) \tag{2.59}$$

Let us now calculate the variance σ^2, again following the same reasoning that led us to equation (2.46):

$$\sigma^2 = \langle S^2 \rangle - \langle S \rangle^2 \; = \int_{-\infty}^{+\infty} df \int_{-\infty}^{+\infty} df' \int_{-\infty}^{+\infty} dk \int_{-\infty}^{+\infty} dk'$$

$$\times \, \delta_T(f - f')\delta_T(k - k')\langle \tilde{n}_1^*(f)\tilde{n}_2(f')\tilde{n}_1^*(k)\tilde{n}_2(k')\rangle \tilde{Q}(f')\tilde{Q}(k')$$

$$= \int_{-\infty}^{+\infty} df \int_{-\infty}^{+\infty} df' \int_{-\infty}^{+\infty} dk \int_{-\infty}^{+\infty} dk' \tag{2.60}$$

$$\times \, \delta_T(f - f')\delta_T(k - k')\langle \tilde{n}_1^*(f)\tilde{n}_1(-k)\rangle \langle \tilde{n}_2^*(-f')\tilde{n}_2(k')\rangle$$

$$\times \, \tilde{Q}(f')\tilde{Q}(k')$$

Substituting equation (2.42) in equation (2.60) we find:

$$\sigma^2 = \frac{1}{4} \int_{-\infty}^{+\infty} df \int_{-\infty}^{+\infty} df' \delta_T^2(f - f')P_1(|f|)P_2(|f'|)\tilde{Q}(f)\tilde{Q}^*(f')$$

$$= \frac{T}{4} \int_{-\infty}^{+\infty} df P_1(|f|)P_2(|f|) \mid \tilde{Q}(f) \mid^2 \tag{2.61}$$

So, to summarise equations (2.59) and (2.60):

$$\mu = \frac{3H_0^2}{20\pi^2} T \int_{-\infty}^{+\infty} df \, |f|^{-3} \, \Omega_{\mathrm{GW}}(|f|)\gamma(|f|)\tilde{Q}(f)$$

$$\sigma^2 = \frac{T}{4} \int_{-\infty}^{+\infty} df P_1(|f|)P_2(|f|) \mid \tilde{Q}(f) \mid^2$$

The problem is now to find the optimal $\tilde{Q}(f)$ that maximises the SNR μ/σ. To this purpose we define an 'inner product' [5] for any pair of complex functions $A(f)$ and $B(f)$:

$$(A, B) \equiv \int_{-\infty}^{+\infty} df A^*(f)B(f)P_1(|f|)P_2(|f|) \tag{2.62}$$

Since $P_{1,2}(|f|) \geqslant 0$, it follows that $(A, A) \geqslant 0$, the equality being valid if and only if $A \equiv 0$. Moreover, $(A, B) = (B, A)^*$ and $(A, B + \lambda C) = (A, B) + \lambda(A, C)$ for any complex number λ. (A, B) is a *positive-definite* inner product that satisfies all the properties of the usual dot product of vectors in three-dimensional Euclidean space.

In terms of this newly defined product, equations (2.59) and (2.60) become:

$$\mu = \frac{3H_0^2}{20\pi^2} T \left(\tilde{Q}(f), \frac{\Omega_{\mathrm{GW}}(|f|)\gamma(|f|)}{|f|^3 P_1(|f|)P_2(|f|)} \right) \tag{2.63}$$

$$\sigma^2 = \frac{T}{4} \left(\tilde{Q}(f), \, \tilde{Q}(f) \right) \tag{2.64}$$

We write the SNR squared:

$$\frac{\mu^2}{\sigma^2} = \left(\frac{3H_0^2}{10\pi^2} \right)^2 T \frac{\left(\tilde{Q}(f), \dfrac{\Omega_{\mathrm{GW}}(|f|)\gamma(|f|)}{|f|^3 \; P_1(|f|)P_2(|f|)} \right)^2}{\left(\tilde{Q}(f), \, \tilde{Q}(f) \right)} \tag{2.65}$$

The best choice for $\tilde{Q}(f)$ is to take it pointing in the same direction as A, so that the cosine of the angle is 1. Therefore,

$$\tilde{Q}(f) = \lambda \frac{\Omega_{GW}(|f|)\gamma(|f|)}{|f|^3 \, P_1(|f|)P_2(|f|)} \tag{2.66}$$

where λ is a real normalisation constant. The important feature of equation (2.66) is that the optimal filter function *depends upon the spectrum* $\Omega_{GW}(f)$, as well as upon the ORF and the power spectra of the detectors. It means that we cannot use a single filter function but rather we need to use a set of filters. For example, within the bandwidth of interest for ground-based detectors such as LIGO and Virgo we look primarily for flat spectra, i.e. $\Omega_{GW}(f) = \Omega_{GW}$ but another very interesting possibility is a power-law dependence [5], such as $\Omega_{GW}(f) = \Omega_a f^\alpha$, as predicted by some cosmological theories [4]. In this case we could construct a set of filters $\tilde{Q}_\alpha(f)$ for some aptly chosen values of α and perform the analysis for each of these filters separately. Equation (2.66) can be rewritten as:

$$\frac{S}{N} = \frac{3H_0^2}{10\pi^2}\left[2T \int_0^\infty df \gamma^2(f) \frac{\Omega_{GW}^2(f)}{f^6 P_1(f)P_2(f)}\right]^{1/2} \tag{2.67}$$

The essential feature of equation (2.67) is that the SNR grows linearly with the observation time T. It means that, at least in principle, for any given noise spectrum of our instruments—a parameter that experimentalists have tried to lower as much as possible since the beginning of interferometric GW detection—and any given strength of the SGWB—an observational parameter—we can let our instruments take data for such a long time as to reach any prescribed SNR.

2.1.5 Statistical considerations

When performing a search for a stochastic background, it is customary to break the data set, which in itself could be months, weeks, days long, into shorter stretches of N points, which we can fast Fourier transform (FFT) and correlate with data from other detectors. Depending on the choice of N and the sampling rate of the detectors, these shorter segments are typically a few seconds long. For example, from equation (2.55) we see that in an observation period of $T = 4$ s the signal S is a sum (over f and f') of approximately 400 independent random variables, since $\tilde{s}_1(f)$ and $\tilde{s}_2(f')$ are correlated only when $|f - f'| < 1/T \approx 0.25$ Hz and the bandwidth over which the integral in equation (2.55) gets its major contribution is roughly $\sim$100 Hz wide. By virtue of the central limit theorem, S is well approximated by a normal random variable. Equivalently we can say that the values of S in a set of measurements over statistically independent time intervals, each of length T, are normally distributed. The mean value of this distribution is $\mu = \langle S \rangle$ and the variance is $\sigma^2 = \langle S^2 \rangle - \langle S \rangle^2$.

Let $\mathbf{s} = (S_1, S_2, ..., S_n)$ be a set of such measurements over statistically independent time intervals, each of length T. Such measurements can be seen as n independent samples drawn from a normal distribution having mean μ and variance

σ^2. The set **s** represents the outcome of a single experiment. From these samples we can construct the sample mean

$$\hat{\mu} = \frac{1}{n} \sum_{i=1}^{n} S_i \qquad (2.68)$$

and the sample variance

$$\hat{\sigma}^2 = \frac{1}{n-1} \sum_{i=1}^{n} (S_i - \hat{\mu})^2 \qquad (2.69)$$

Given the values of these estimators, we can decide whether or not we have detected a stochastic signal. To make such a decision, we use a standard theorem from conventional statistics and probability. The theorem states that if $\hat{\mu}$ is the sample mean of n independent samples with mean μ and variance σ^2 then

$$t = \frac{\hat{\mu} - \mu}{\sigma/\sqrt{n}} \qquad (2.70)$$

is the value of a random Student variable, i.e. a variable that follows the Student distribution with parameter $\nu = n - 1$. The Student's distribution is a standard distribution, whose values can be found in all handbooks on statistics, so it is handy to use. The drawback is that the calculations depend on the parameter $\nu = n - 1$. So every result on the SGWB would in principle depend on n, the number of individual observations which make up a single experiment. To overcome this problem, we notice that for large n the Student's distribution is almost indistinguishable from the normal distribution. Since a typical total observation time will be of the order of years ($\sim 10^7$ s) and T is of the order of a few seconds, as stated above, it is no problem to choose $n \sim 10^3$. For such large n, equation (2.70) can be written as

$$z \simeq \frac{\hat{\mu} - \mu}{\hat{\sigma}/\sqrt{n}} \qquad (2.71)$$

where z is a normal variable with zero mean and unit variance.

2.1.6 Signal detection

Given a set of data points, we thus need a rule to tell from two alternative and mutually exclusive hypotheses:

H_0: no SGWB signal is present.

H_1: there is a signal in our data, with *unknown* mean $\mu > 0$.

H_0 is a simple hypothesis, since it does not depend on any parameter. H_1 is a composite hypothesis since it depends on a range of the unknown parameter μ, explicitly

$$H_1 = \bigcup_{n=0}^{\infty} H_{nd\mu, d\mu} \qquad (2.72)$$

Hypothesis testing, both Bayesian and frequentist, is the subject of many books and articles, so we won't explore things in detail. Let us just state the detection strategy. Let $\mathbf{s} = (S_1, S_2, ..., S_n)$ be a set of cross-correlation measurements. $\mathbf{s}$ is a random variable characterised by the probability density functions:

$$p(\mathbf{s}|0) = (2\pi\hat{\sigma}^2)^{-n/2} \exp\left[-\sum_{i=1}^{n} \frac{S_i^2}{2\hat{\sigma}^2} \right] \tag{2.73}$$

which is the probability (density of function) of $\mathbf{s}$ as an outcome given that no signal is present, and

$$p(\mathbf{s}|\mu) = (2\pi\hat{\sigma}^2)^{-n/2} \exp\left[-\sum_{i=1}^{n} \frac{(\mu - S_i)^2}{2\hat{\sigma}^2} \right] \tag{2.74}$$

which is the probability of finding $\mathbf{s}$ with a signal of given (but unknown) mean μ.

Finding a rule that selects either H_0 or H_1 means that we divide the space of all possible experimental outcomes in two non overlapping regions R_0 and R_1. If $\mathbf{s} \in R_0$ then H_0 is chosen, if $\mathbf{s} \in R_1$ then we choose H_1. Associated with this rule are two errors: the false alarm, i.e. the error we make when we choose H_1 but no signal is present, and the false dismissal, i.e. the error we make when we choose H_0 in the presence of a signal. In formulas, the false alarm rate α is given by:

$$\alpha = \int_{R_1} d\mathbf{s}\, p(\mathbf{s}|0) \tag{2.75}$$

and the false dismissal rate β is:

$$\beta(\mu) = \int_{R_0} d\mathbf{s}\, p(\mathbf{s}|\mu) \tag{2.76}$$

Since H_1 is a complex hypothesis, the false dismissal rate is actually a function of the unknown parameter μ. It is remarkable that $1 - \alpha$ is nothing but the fraction of experimental outcomes where the decision rules correctly identify the absence of the signal. In terms of detection, we introduce the detection rate γ as:

$$\gamma = 1 - \beta(\mu) \tag{2.77}$$

as the fraction of experimental outcomes where the decision rules correctly identify the presence of a signal. The optimal choise of the two regions R_0 and R_1 is called in the literature the criterion of Newman–Pearson. We simply state the result: we choose H_0 if $\hat{\mu} < z_a \hat{\sigma}/n$ and conversely we choose H_1 if $\hat{\mu} \geqslant z_a \hat{\sigma}/n$ where z_a is that value of the random variable for which the area of the standard normal distribution ($\mu = 0$, $\sigma^2 = 1$) to its right is equal to α. This decision rule can be restated as: choose H_0 if $n\hat{\mu}/\hat{\sigma} = n\,\widehat{\text{SNR}} < z_a$ and choose H_1 if $n\,\widehat{\text{SNR}} \geqslant z_a$. We stress once more the importance of the SNR. A similar approach can be followed once we have decided that there is a stochastic signal and we want to calculate μ. In this case

$$\hat{\mu} - z_{\alpha/2}\hat{\sigma}/n < \mu < \hat{\mu} + z_{\alpha/2}\hat{\sigma}/n \qquad (2.78)$$

where in this case $z_{\alpha/2}$ is the value of the random variable for which the standard distribution area to its right equals $\alpha/2$. For more details we refer the reader to the already mentioned article by Allen and Romano [5], which stands as a reference for most of this chapter, as well as to standard texts on statistics such as [6] and [7].

2.1.7 Observational bounds

Before diving deep into the theory of sources, we close this section by discussing what we already know about the SGWB from an observational side. Two bounds we have that are well outside the frequencies investigated by our detectors are coming from timing irregularities in the arrival time of the pulses from some millisecond pulsars and from large scale anisotropies of the CMB, respectively. Millisecond pulsar observations constrain Ω_{GW} in the frequency region $f \sim 10^{-8}$ Hz. Actually, millisecond pulsars can be used to detect GWs albeit indirectly [8]. The other bound puts some constraints in the 10^{-18} Hz $\leq f \leq 10^{-16}$ Hz. They are important results but of no interest to our current experimental efforts.

By far the most stringent bound comes from Big Bang nucleosynthesis (BBN). We will follow [9]. Nucleosynthesis happened when the Universe was $t = 180$ s old. Nucleosynthesis theory correctly predicts the primordial abundances of light elements such as deuterium, ^{3}He, ^{4}He and ^{7}Li in terms of a single cosmological parameter, the baryon-to-photon ratio η, which is observationally given by $\eta \sim 5 \times 10^{-10}$ [2]. Other theory-dependent parameters enter the calculations too and are thus required not to spoil the excellent agreement between observation and theory. In particular the prediction is very sensitive to the *effective number of species* at the time of nucleosynthesis, $g_* = g(T \sim 1$ MeV$)$. This dependence can be understood as follows: a crucial parameter in the nucleosynthesis computation is the density of neutrons to the density of protons ratio, n_n/n_p. So long as thermal equilibrium holds, $T > 800$ keV, we have $n_n/n_p = \exp(-Q/T)$ where $Q = m_n - m_p = 1.29$ MeV. Equilibrium is physically maintained by the $pe \longleftrightarrow n\nu$ process, at least as its width is bigger than the Hubble rate. When the width drops below the Hubble constant, the process cannot compete with the expansion of the Universe, equilibrium is lost. Free neutrons decay until $T \sim 60$ keV, at which time

$$\frac{n_n}{n_p}(60 \text{ keV}) \sim 0.2 \exp(-\Delta t/\tau_n) \sim 0.1 \qquad (2.79)$$

where $\Delta t = t(60$ keV$) - t(800$ keV$) \sim 3$ min and τ_n is the decay time for neutrons.

At $T \sim 60$ keV the remaining neutrons are rapidly incorporated into nuclei via a series of reactions, and since practically all the neutrons form ^{4}He, the final primordial abundance of this element is extremely sensitive to T_f. If we assume for the sake of simplicity $\Gamma_{pe \longleftrightarrow n\nu} \simeq G_F^2 T^5$ which is not a rigorous argument but is valid when $T \gg Q$, then

$$G_F^2 T^5 \simeq \left(\frac{4\pi g_*}{45}\right)^{1/2} \frac{T_f^2}{M_{Pl}} \qquad (2.80)$$

This shows that $T_f \propto g_*^{1/6}$. A large energy density in relic gravitons would give a large contribution to the total energy density ρ and therefore to g_*. This in turn would result in a larger freeze-out temperature, more available neutrons and in a larger than observed production of ^{4}He. This is the idea behind the nucleosynthesis bound. In itself, a larger g_* could be compensated by a decrease of η, but this would affect the abundance of deuterium and ^{3}He, and this is not what we observe.

Rather than g_*, an effective number of neutrino species, N_ν is often used, defined as follows. In the Standard Cosmological Model, when the temperature is still of the order of a few MeV, the active degrees of freedom are the photon, $e^\pm$, neutrinos as well as antineutrinos and they have the same temperature. Then for N_ν families of light neutrinos, we have:

$$g_*(N_\nu) = 2 + \frac{7}{8}(4 + 2N_\nu) \tag{2.81}$$

where the factor 2 counts the elicity states of the photon—which is its own antiparticle—the factor 4 comes from the two polarisation states of the electron and of the antielectron and $2N_\nu$ counts the N_ν neutrinos and the N_ν antineutrinos, each with its single elicity state. According to the Standard Model of elementary particles, $N_\nu = 3$ and therefore $g_* = \frac{43}{4}$. Therefore, we can define an 'effective number of neutrino species' as:

$$g_*(N_\nu) = \frac{43}{4} + \sum_{i=\text{extra bosons}} g_i\left(\frac{T_i}{T}\right)^4 + \sum_{i=\text{extra fermions}} g_i\left(\frac{T_i}{T}\right)^4 \tag{2.82}$$

An extra species of light neutrino, in thermal equilibrium with the photons, would contribute one unit to N_ν but all species do contribute to N_ν, each with itw own energy density. N_ν need not be an integer. For i = gravitons, we have $g_i = 2$ and $(T_i/T)^4 = \rho_{GW}/\rho_\gamma$ where $\rho_\gamma = 2(\pi^2/30)T^4$ is the photon energy density. If GWs are the only extra contribution to equation (2.82) we find immediately

$$\left(\frac{\rho_{GW}}{\rho_\gamma}\right)_{NS} = \frac{7}{8}(N_\nu - 3) \tag{2.83}$$

where the subscript means that this equations holds only for extra degrees of freedom already present at nucleosynthesis time. So this bound does not apply to a stochastic foreground of astrophysical origin. If more extra species, not included in the Standard Model, happen to contribute, the equality in equation (2.83) must be replaced with less than or equal ($\leq$). Such contributions could come, for example, from primordial black holes.

This bound, as already stated, holds at nucleosynthesis time. To translate it to present time, we must scale the relevant quantities by the scale factors due to expansion of the Universe:

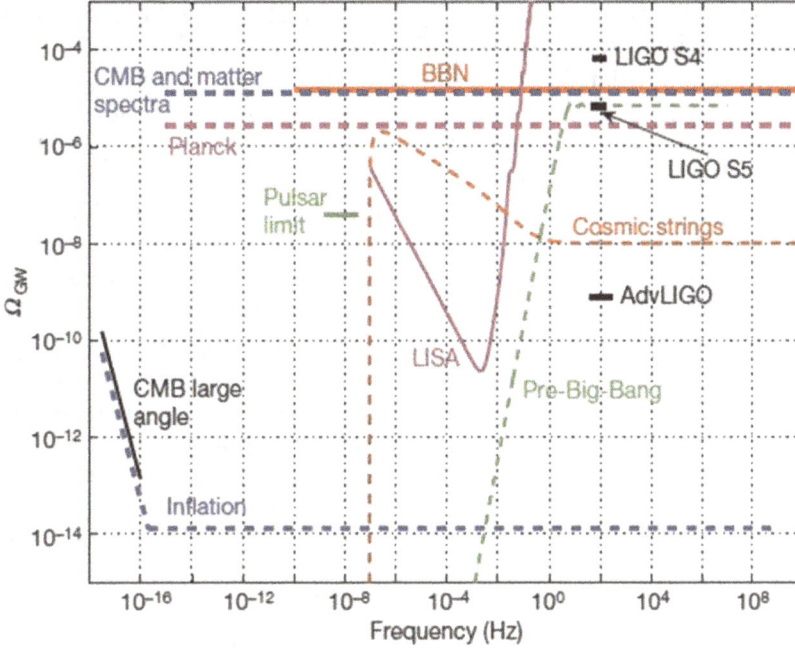

Figure 2.3. Comparison of different SGWB measurements and models.

$$\left(\frac{\rho_{GW}}{\rho_\gamma}\right)_0 = \left(\frac{\rho_{GW}}{\rho_\gamma}\right)_{NS}\left(\frac{g_S(T_0)}{g_S(1\ \text{MeV})}\right)^{4/3} = \left(\frac{\rho_{GW}}{\rho_\gamma}\right)_{NS}\left(\frac{3.91}{10.75}\right)^{4/3} \tag{2.84}$$

therefore we get the nucleosynthesis bound at present time:

$$\left(\frac{\rho_{GW}}{\rho_\gamma}\right)_0 \leq 0.227(N_\nu - 3) \tag{2.85}$$

This is a bound on the *total energy density* in GWs, integrated over all frequencies. We can rewrite it as:

$$\int_{f=0}^{f=+\infty} d(\ln f)h_0^2\Omega_{GW}(f) \leq 5.6 \times 10^{-6}(N_\nu - 3) \tag{2.86}$$

This analysis suffers from a few systematic errors. A very conservative limit is $N_\nu < 4$. A less conservative limit is $N_\nu < 5$. Correspondingly, the r.h.s of equation (2.86) is conservatively of the order 5×10^{-6} and cannot be larger than 10^{-5}. If the integral cannot exceed these values, neither can the integrand, which is positive defined, over an appreciable interval of frequencies, $\Delta \ln f \sim 1$. One might still have, in principle, a very narrow peak in $h_0^2\Omega_{GW}(f)$ for some frequency f, with peak value larger than 10^{-5}, while its contribution to the integral could be small enough. However, this scenario is implausible and not suggested by any cosmological model.

We therefore often translate the bound given by equation (2.86) to a bound over the integrand, assuming a spectrum flat in frequencies.

The article published in 2009 by the LIGO Virgo Collaboration (LVC) [10] was a fantastic achievement and set stringent upper limits on the SGWB based on observational GW data itself rather than on indirect evidence. Its publication was probably LVC's biggest and most significant achievement before detection. It was claimed that the new observational bound surpassed the BBN bound. However this claim must be taken with caution, as mentioned in the previous paragraph. The comparison between the GW data and the BBN bound is only meaningful if we assume for the SGWB a flat spectrum, independent of frequency. Figure 2.3 and the caption are taken from that article. We remark moreover that the BBN bound applies only to those sources of cosmological origin that were already present at nucleosynthesis time ($t = 180$ s), whereas the observational bound applies to all stochastic sources of cosmological and of astrophysical origin.

A recent paper [11] has improved the upper bound by an order of magnitude in LIGO's most sensitive frequency band: 24 Hz to 80 Hz. The new upper bound for a flat spectrum is

$$h_0^2 \Omega_{\mathrm{GW}} < 1.7 \times 10^{-7} \tag{2.87}$$

with 95% confidence. The same article sets upper bounds also for some power-law spectra which could be of importance in some cosmological models. A power-law spectrum is a spectrum of the form

$$\Omega_{\mathrm{GW}}(f) = \Omega_\alpha \left(\frac{f}{f_{\mathrm{ref}}} \right)^\alpha \tag{2.88}$$

and the article deals extensively with the cases $\alpha = 2/3$ and $\alpha = 3$.

References

[1] Hartle James B 2003 *Gravity: An Introduction to Einstein's General Relativity* (Reading, MA (also London): Addison Wesley)

[2] James Rich 2001 *Fundamentals of Cosmology* (Berlin: Springer)

[3] Penzias A A and Wilson R W 1965 A measurement of excess antenna temperature at 4080 Mc s^{-1} *Astrophys. J. Lett.* **142** 419–21

[4] Maggiore Michele 2008 *Gravitational Waves. Volume 1: Theory and Experiments* (Oxford: Oxford University Press)

[5] Allen Bruce and Romano Joseph D Mar 1999 Detecting a stochastic background of gravitational radiation: Signal processing strategies and sensitivities *Phys. Rev. D* **59** 102001

[6] Kay Steven M 1993 *Fundamentals of Statistical Signal Processing: Estimation theory* (Englewood Cliffs, NJ : Prentice Hall)

[7] Kay Steven M 1998 *Fundamentals of Statistical Signal Processing: Detection theory* (Englewood Cliffs, NJ: Prentice Hall)

[8] Sesana A, Vecchio A and Colacino C N 2008 The stochastic gravitational-wave background from massive black hole binary systems: implications for observations with pulsar timing arrays *Mon. Not. R. Astron. Soc.* **390** 192–209

[9] Losurdo G, Maggiore M, Matone G, Babusci D, Foffa S and Sturani R 2001 The stochastic gravitational-wave background. *Gravitational Waves* eds I Ciufolini, V Gorini, U Moschella and P Fre (Boca Raton, FL: CRC Press) p181

[10] Abbott B P *et al* 2009 An upper limit on the stochastic gravitational-wave background of cosmological origin *Nature* **460** 990–4

[11] Abbott B P *et al* (LIGO Scientific Collaboration and Virgo Collaboration) 2017 Upper limits on the stochastic gravitational-wave background from Advanced LIGO's first observing run *Phys. Rev. Lett.* **118** 121101

[12] Adam R *et al* (The Planck Collaboration) 2017 Planck intermediate results. XXX. The angular power spectrum of polarized dust emission at intermediate and high galactic latitudes *Astron. Astrophys.* **586** 121101

Chapter 3

SGWB generation and source detection

As mentioned in the introduction, an SGWB component of cosmological origin could shed light on some of the most profound and disturbing mysteries of the very early Universe. Let us now face the most relevant questions: how was an SGWB generated? What are the sources for which we are looking? What are the sources we can detect with our instruments? Are the answers to these two questions really the same? Some of the sources are even mentioned in the caption of figure 2.3. We will follow once again the treatment by Michele Maggiore [1], integrating it with more recent results, because both Michele Maggiore and myself are trained high-energy physics and our way of studying sources slightly differs from that of astrophysicists. Much of the following material is purely speculative, the fact is, we don't know much about the very early Universe and GWs could be a fantastic probe to remove the veil of our ignorance! This is however a list of some processes that might have produced a primordial GW spectrum.

3.1 Topological defects

The concept of spontaneous symmetry breaking is one of the most important in particle physics and quantum field theory [2]. The whole Standard Model of Particle Physics is constructed upon gauge symmetries and their spontaneous breaking, of which the Higgs field, i.e. the field that gives every particle its mass, is the most spectacular manifestation [3]. The distinction between spontaneous and explicit symmetry breaking is that with spontaneous symmetry breaking the Lagrangian is invariant under the symmetry but the ground state of the theory (the vacuum) is not. With explicit symmetry breaking there was never an exact symmetry to begin with. Ferromagnetic materials are the most ubiquitous example of spontaneous symmetry breaking. Spontaneous symmetry breaking is associated with phase transitions as the temperature is lowered. And phase transitions are a crucial feature of standard Big Bang cosmology, with critical temperatures related to symmetry breaking scales [4].

doi:10.1088/978-1-6817-4082-9ch3

To illustrate the idea of spontaneous symmetry breaking and to see how it works in standard cosmology, let us consider a single scalar field with Lagrangian:

$$\mathcal{L} = \frac{1}{2}(\partial_\mu \phi)^2 - \frac{1}{2}m^2\phi^2 - \frac{\lambda}{4!}\phi^4 \tag{3.1}$$

In analogy with ferromagnetism, let's assume there is a critical temperature T_c for which there is symmetry breaking. Near T_c we can expand the mass term $m^2(T) = c(T - T_c)$. For $T > T_c$ the rhs of this equation is positive and the mass term has the right sign, i.e. $m^2 > 0$. The Lagrangian describes an ordinary field theory. For $T < T_c$ however, $m^2 < 0$. Then the extremum $\phi = 0$ is a local maximum of the potential $V = -\mathcal{L}_{\text{int}} = \frac{1}{2}m^2\phi^2 + \frac{\lambda}{4!}\phi^4$ instead of a minimum. Furthermore, it is unstable.

For $T < T_c$ we replace $m^2 \to - m^2$ so the mass term is still positive and we have:

$$\mathcal{L} = \frac{1}{2}(\partial_\mu \phi)^2 + \frac{1}{2}m\phi^2 - \frac{\lambda}{4!}\phi^4 \tag{3.2}$$

This Lagrangian has a Z_2 symmetry under $\phi \to - \phi$. There are two possible minima of the potential, $\phi = \pm\sqrt{\frac{6m^2}{\lambda}}$. There are two possible different vacua: $|\Omega_+ \rangle$ with

$$\langle \Omega_+ |\phi|\Omega_+ \rangle = \sqrt{\frac{6m^2}{\lambda}} \tag{3.3}$$

and $|\Omega_- \rangle$ with

$$\langle \Omega_- |\phi|\Omega_- \rangle = -\sqrt{\frac{6m^2}{\lambda}} \tag{3.4}$$

At either minimum, Z_2 is spontaneously broken because it takes $|\Omega_+ \rangle \leftrightarrow |\Omega_- \rangle$. The two possible vacua for the theory are in principle equivalent but one has to be chosen. The vacuum breaks the symmetry, which remains in the Lagrangian though [2].

In a cosmological contest, the Universe cools continually. As it passes the critical temperature T_c the Higgs field ϕ develops an expectation value $\langle\phi\rangle$ corresponding to some point in the manifold $\mathcal{M}$. All points in the manifold are equivalent, as seen, the choice depends on random fluctuations that are different in different regions of spaces. Associated with the phase transition there is a correlation length ξ representing the maximum length over which the Higgs field can be correlated. We have

$$\xi \leqslant t_c \tag{3.5}$$

where t_c is the time at which the phase transition happened. On the boundary between different correlation regions ϕ assumes values that minimise the potential. For topological reasons these regions of false vacuum are stable and do not change in the evolution of the Universe, but survive as *topological defects* [5]. To illustrate that, let's take for example the phase transition known as 'freezing water'. At temperatures $T > 273$ K water is liquid. Individual water molecules are randomly oriented and the liquid water has thus rotational symmetry about any point. It is isoptropic. However, when the water cools below $T = 273$ K the water undergoes a phase transition from liquid to solid and the rotational symmetry is lost. The water molecules are locked into a crystalline structure and the ice no longer has rotational symmetry about an arbitrary point. The ice crystal is anisotropic, with preferred directions corresponding to the crystal's axes of symmetry. To see how topological defects arise, let's consider a large tub of water cooled below the freezing point. Usually the freezing will start at two or more widely separated sites. The crystal that forms about any given nucleation site is very regular, with well defined axes of symmetry. However, the axes of symmetry of two adjacent ice crystals need not be aligned. At their boundary there will be a two-dimensional topological defect, called *a domain wall*, where the axes of symmetry fail to line up. In cosmology this production of topological defects is known as *Kibble mechanism* [6]. The analogies between defects in particle physics and in condensed matter physics is quite deep. However their dynamics could not be more different: the dynamics in condensed matter is friction dominated, whereas in cosmology defects obey a second order differential equation, with boundary conditions dictated by the underlying particle theory.

Depending on the topology of the manifold $\mathcal{M}$, defects can occur as points (0 dimensions), one-dimensional lines, or surfaces. They are called (magnetic) monopoles, (cosmic) strings and domain walls, respectively. Domain walls and magnetic monopoles are an unwelcome feature in cosmological theories, their energy density would soon come to dominate the Universe and are therefore ruled out [7]. Even in some grand unification theories (GUTs) magnetic monopoles are not produced. Cosmic strings on the other hand can lead to very interesting comological consequences and have been widely investigated both in the framework of structure formation and for gravitational radiation [8].

3.2 Strings

Strings are associated with the spontaneous breaking of a local $U(1)$ symmetry. In this case the Lagrangian contains a gauge field A^μ and a complex Higgs field ϕ which carries $U(1)$ charge g and with self interaction of the form:

$$\mathcal{L} = (D_\mu \phi)^\dagger (D^\mu \phi) - \frac{1}{4} F_{\mu\nu} F^{\mu\nu} - V(\phi) \tag{3.6}$$

where

$$D_\mu \equiv \partial_\mu - ig A_\mu \qquad F_{\mu\nu} \equiv \partial_\mu A_\nu - \partial_\nu A_\mu$$

Applied to strings, the Kibble mechanism implies that at time t_c a network of strings will form, whose typical length is $\xi(t_c)$. The spectrum of GW radiation produced by a string network has been calculated by [9]. A loop radiates with power

$$P = \Gamma G \mu^2 \tag{3.7}$$

where Γ is a dimensionless constant of order 60, G (or G_N) is Newton's constant, and μ is the mass per unit length of the string. The spectrum has two main features:

- A nearly equal gravitational radiation energy density per logarithmic frequency interval in the range $(10^{-8}, 10^{10})$ Hz. This portion is called *red noise* and corresponds to GWs emitted during the radiation-dominated era.
- A peak near $f \sim 10^{-12}$ Hz. The shape of this portion depends critically on the details of the model.

The red noise is accessible to ground-based detectors and we will focus on it. The energy density is given by

$$\Omega_{GW}(f) = \frac{8\pi}{9} \frac{A\Gamma G^2\mu^2}{\alpha}(1 - \langle v^2\rangle)\frac{\beta^{3/2} - 1}{1 + z_{eq}} \tag{3.8}$$

where A and v^2 are dimensionless constants, α is the fixed ratio between the size of a newly formed loop and the horizon, $l = \alpha d_H(t)$, and β is the parameter that relates the time t_d, time in which the length loop goes to zero, to t_b, time the loop was formed: $t_d = \beta t_b$. Numerical simulation provides $A = 52 \pm 10$ and $\langle v^2\rangle = 0.43 \pm 0.02$ in the radiation-dominated era. From CMB temperature anisotropies [7] we get:

$$G\mu = 1.05^{+0.35}_{-0.20} \times 10^{-6} \tag{3.9}$$

Unfortunately, the value of α—the size of the loop at its formation—is still uncertain. Numerical simulations seem to imply that $\alpha < 10^{-2}$. This uncertainty is obviously reflected on β, the parameter which governs the lifetime of the loop. Conservatively we get:

$$\Omega_{GW} \geqslant 1.6 \times 10^{-9} \tag{3.10}$$

3.3 Inflation

Although it is safe to say that the theory of inflation is still looking for experimental confirmation, the inflationary paradigm is now an accepted part of the Standard Cosmological Model. This paradigm basically states that, before radiation, during which the Friedmann–Robertson–Walker scale factor $a(t)$ was growing as $\sqrt{t}$, there was an earlier period, called inflation, when the energy density of the Universe was dominated by a slow varying vacuum energy, and $a(t)$ grew more or less exponentially. There are many different inflationary models—I once said myself at a conference only half jokingly that I had counted no less than 43 different inflationary models in the literature—but the mechanism of inflation is recognised as an essential part of the history of the Universe because it solves some of the more intricate mysteries of the standard Big Bang model, such as the horizon and the flatness problem [5]. Besides, inflation has conquered even the most sceptical cosmologists in that it seems to predict some of the properties of the fluctuations of the cosmic microwave background and large scale structure [8]. Inflation, as we said before, can be defined as the hypothesis that there was a period when the expansion was accelerating outwards. Friedmann's acceleration equation is

$$\frac{\ddot{a}}{a} = -\frac{4\pi G}{3}(\varrho + 3p) \qquad (3.11)$$

where a is the scale factor in the Friedmann–Roberston–Walker Universe, ϱ is the energy density of the Universe and p the pressure. A positive acceleration implies $\varrho + 3p < 0$, i.e. $p < -\varrho/3$. Thus for inflation to take place, the Universe had to be temporarily dominated by a component with equation of state parameter $w = \frac{p}{\varrho} < -\frac{1}{3}$. The usual implementation of this states that the Universe was dominated by a *positive cosmological constant* Λ, and thus had an acceleration equation written in the form:

$$\frac{\ddot{a}}{a} = \frac{\Lambda}{3} > 0 \qquad (3.12)$$

In the inflationary phase the Friedmann equation was:

$$\left(\frac{\dot{a}}{a}\right)^2 = \frac{\Lambda}{3} \qquad (3.13)$$

The Hubble constant during the inflationary phase was thus constant also in time. $H_i = (\Lambda/3)^{1/2}$ and the scale factor grew exponentially as

$$a(t) \propto \exp(H_i t) \qquad (3.14)$$

This exponential growth solves the flatness, horizon and monopole problems of the Universe. Moreover, it takes the submicroscopic quantum fluctuations in the inflaton field, be they scalar (density) or tensor (GWs) fluctuations, and expands them to macroscopic scales. The basic idea is quite simple: zero point quantum fluctuations of a field can be amplified when the size of the perturbation becomes larger than the event horizon. This is analogous to the Unruh effect, for which an accelerated observer sees a thermal bath of particles in a state which, seen from an observer at rest, is the vacuum state [10]. The energy fluctuations, due to quantum fluctuations of the vacuum, are the origin, in the inflationary scenario, of the inhomogeneities in the present Universe. Before investigating what kind of GW wave spectrum we can expect from inflation, let us recall that a few years ago the BICEP2 collaboration announced the detection of a B-polarisation component in the CMB spectrum [11]. A B-polarisation component of primordial origin is generated at decoupling by Thomson scattering if the quadrupole temperature anisotropy of the photons, caused by metric perturbations, has a tensor component. That would definitely be a signature of primordial GWs from inflation. A few months later, this enthusiasm was heavily damped when new observations from Planck showed that the BICEP2 signal was strongly contaminated by galactic dust [12]. That in itself does not rule out a primordial GW signal but it shows that the foreground must be taken carefully into account in the analysis of the data. The detection of GWs from inflation would have a major impact on physics because on one hand it would mean that gravity can be quantised the same way as any other classical field, at least in the linear approximation, and on the other hand it would set the value for the energy scale of inflation in the simplest slow-roll model, $V^{1/4} \simeq 10^{16}$ GeV. Discrepancies from this value would reveal further details about the inflationary model.

3.3.1 The inflationary spectrum

As we mentioned earlier, inflation is more a paradigm than a sound model. In this section we outline some of the ideas that have been put to the fore for the calculation of the spectrum of primordial GWs generated by the inflationary amplification of fluctuations.

The starting point is the usual separation of the metric into a static background and a propagating part:

$$g_{\mu\nu} = a^2(\eta)(-\mathrm{d}\eta^2 + \mathrm{d}\vec{x}^2) + h_{\mu\nu} \tag{3.15}$$

where a is the usual scale factor of the Universe, which depends only on the conformal time η, $\mathrm{d}\eta = \mathrm{d}t/a(\eta)$. $h_{\mu\nu}$ can be expanded in plane waves as:

$$h_{\mu\nu} = \frac{1}{a(\eta)} e_{\mu\nu}(\vec{k}) \psi(\eta) \exp(i\vec{k}\cdot\vec{x}) \tag{3.16}$$

where $e_{\mu\nu}(\vec{k})$ is the polarisation tensor. This is similar to what we did in a previous section, see equation (1.17). The equation of motion is:

$$\psi''(\eta) + \left(k^2 - \frac{a''(\eta)}{a(\eta)} \right) \psi(\eta) = 0 \tag{3.17}$$

where the derivatives are with respect to conformal time. This is a Schrödinger equation with potential given by:

$$V(\eta) = \frac{a''(\eta)}{a(\eta)} \tag{3.18}$$

If $k \gg V(\eta)$, then we have the obvious solution $\psi \sim \exp(-ik\eta)$, hence $|h_{\mu\nu}| \sim a^{-1}(\eta)$; if instead $k \ll V(\eta)$ we get two solutions:

$$\psi_1 \sim a(\eta) \qquad \psi_2 \sim a(\eta) \int \frac{\mathrm{d}\eta}{a^2(\eta)}$$

ψ_1 is the dominant solution, in this case $|h_{\mu\nu}| \sim 1$. This means that a solution with a long wavelength, i.e. one that obeys the condition $k \ll V(\eta)$, is amplified with respect to a short wavelengthed one by a factor $a(\eta)$; as a first approximation we can aasume this amplificaton to happen as long as $k < V(\eta)$. This inequality can be recast, apart from numerical factors, as:

$$\frac{k}{a} < \frac{a'(\eta)}{a^2(\eta)} \equiv H(\eta) \tag{3.19}$$

Equation (3.19) relates the physical momentum of the perturbation to the Hubble parameter, whose inverse corresponds to the horizon size. This is, as already mentioned, the key feature of inflation: a perturbation is amplified when its wavelength becomes larger than the event horizon. From the observed anisotropies in the CMBR the spectrum is constrained as follows:

$$\Omega_{GW} \leqslant 7 \times 10^{-11} \left(\frac{H_0}{f}\right)^2 \quad \text{for} \quad f \in (10^{-19}, 10^{-16})\text{Hz} \tag{3.20}$$

In the frequency range probed by LIGO and Virgo, assuming the simplest model of inflation we find:

$$\Omega_{GW} \leqslant 8 \times 10^{-14} \tag{3.21}$$

which is well below the sensitivity of current detectors, but is the target of future experiments such as DECIGO and BBO.

3.3.2 Particle production during inflation

In addition to the quantum amplifications of vacuum fluctuations and of the amplifications of long wavelength perturbations, GWs can be produced classically if anisotropic stress is present. Anisotropic stresses can be generated by particles: in slow-roll inflation, as the inflaton rolls towards its potential minimum, it provides a time-changing effective mass to fields coupled to it. Should one of these fields become massless during inflation, quanta of this field, i.e. particles, could be generated in a non-perturbative way. The most favourable scenario is the coupling between the inflaton and a gauge field of the form $\phi F_{\mu\nu}\tilde{F}^{\mu\nu}$; such a coupling is natural in models where the inflaton is an axion. Such a mechanism could give rise to a GW spectrum detectable by ground-based detectors. These models however are severely constrained by CMB non Gaussianities and have little room for additional parameters. On the other hand, there is a widespread interest for models where the inflaton is an axion, since axions are becoming fashionable in the contest of dark matter theories.

3.3.3 Preheating

GW production can occur also at the end of inflation when the potential energy density that had given rise to inflation is converted into matter and above all radiation energy density. This process is called reheating. In many inflationary models reheating starts with preheating, a non-perturbative decay of the inflaton into fluctuations of itself and of the other fields coupled to it. The system evolves towards thermodynamical equilibrium in a highly non-linear and turbulent way. Large fluctuations of the fields give rise to anisotropic stress which, as stated above, can produce gravitational radiation. Preheating is a causal process, GWs are emitted at a characteristic wave number $k_* = H_*/\varepsilon_*$ where H_* is the Hubble constant at the end of inflation and ε_* depends on the details of the inflationary model. The inflationary energy scale must be smaller than 10^{11} GeV for the peak of the spectrum to fall into the frequency range probed by ground-based detectors and no larger than 10^7 GeV to be detected by eLISA. Gws from preheating would thus be observable with current ground-based detectors if inflation had happened at relatively low-energy [13], so such an observation would tell us a lot about inflation. On the other hand, if the observation from BICEP2 proved to be of cosmological origin and not merely due to intergalactic dust, a low-energy inflation would be ruled out, and no GWs from preheating would be detectable.

3.4 String cosmology

String cosmology is the attempt to reconcile string theory with cosmology. We refer the interested reader to the book by Maurizio Gasperini [14]. The attractive feature of this model is that the spectrum of gravitational radiation grows with frequency, thus ground-based detectors such as LIGO and Virgo could detect it. In string theory the fundamental objects are one-dimensional extended entities called strings. Their fundamental excitations of given energy and angular momentum correspond to particles with given mass and spin. Although there are plenty of string theories in literature, it is thought that those different theories are just branches of a fundamental underlying theory, which has only one fundamental constant: *the string tension* T which can be traded for a fundamental string length λ_s: $\lambda_s \equiv \sqrt{\frac{\hbar c}{T}}$.

The mass scale of the excitation of the string is therefore $\sqrt{T}$ which, since string theory claims to be a theory of quantum gravity, among other things, should be of the order of the Planck mass. The gauge couplings are not fixed, they are dynamical and depend on the expectation value of a fundamental scalar field, the *dilaton*. For example the Newton gravitational constant is given by:

$$G_N \simeq \frac{2\lambda_s}{8\pi} \exp(\phi) \tag{3.22}$$

and other gauge couplings are proportional to $\exp(\phi/2)$. At low energy string theory is adequately described by an effective field theory whose action is:

$$S = -\frac{1}{2\lambda_s^2} \int d^4x \sqrt{-g} [\exp(-\phi)(R + \partial_\mu\phi\partial^\mu\phi) + V(\phi)] + B \tag{3.23}$$

where $g \equiv \det \|g_{\mu\nu}\|$, R is the Ricci scalar, $V(\phi)$ is a potential, and B stands for higher order terms in $\exp(\phi)$, which clearly must be taken into account in the full quantum regime. The terms without corrections are nothing but the Einstenian gravity with a constant scalar field (the dilaton).

Let us study a spatially homogenous field, $\phi = \phi(t)$ and the metric is chosen so that the line element ds^2 can be written as:

$$ds^2 = -dt^2 + a^2(t)d\vec{x}^2 \tag{3.24}$$

The cosmological field equations which stem from equation (3.23) are symmetric under the transformations

$$a(t)\frac{1}{a(-t)}, \qquad \phi(t)\phi(-t) - 6\ln[a(-t)] \tag{3.25}$$

which relate ordinary Friedmann–Robertson–Walker cosmology characterised by $H = \dot{a}/a > 0$, $\dot{H} < 0$ and constant ϕ at $t > 0$ with an inflationary one with $H > 0$,

$\dot{H} > 0$ and $\dot{\phi} > 0$ at $t < 0$. The dual cosmology is called the pre-Big Bang (PBB) phase. The problem with this scenario is that low-energy equations of motion do not smoothly interpolate between these two phases but rather lead to singularities, which could be avoided only in the full string quantum regime.

In order to calculate the spectrum, some assumptions must be made. We use a simple model of background which undergoes two transitions, a first one from inflation to radiation-dominated expansion at $\eta = -\eta_1$ and a second transition at $\eta = \eta_{eq}$ to the final matter-dominated regime. η is as usual the conformal time. We also introduce a new parameter, the proper frequency $\omega(t) = k/a(t)$. Furthermore, we assume the dilaton field to be dynamical during inflation and to become frozen, trapped at a minimum of its potential, during the subsequent epochs. The resulting energy density is given by:

$$
\begin{aligned}
\Omega_{GW}(\omega, t_0) &= \left(\frac{H_1}{M_{Pl}}\right)\Omega_r(t_0)\left(\frac{\omega}{\omega_1}\right)^{2+2\alpha} \qquad \omega_{eq} < \omega < \omega_1 \\
&= \left(\frac{H_1}{M_{Pl}}\right)^2 \Omega_r(t_0)\left(\frac{\omega}{\omega_1}\right)^{2+2\alpha}\left(\frac{\omega_{eq}}{\omega}\right)^2 \qquad \omega_0 < \omega < \omega_{eq}
\end{aligned}
\tag{3.26}
$$

There are three important parameters, closely related to inflation: I) the curvature scale H_1, which controls the amplitude of the gravitational radiation background; II) the kinematic power α, which controls the slope of the spectrum and which is thought to satisfy $\alpha < 1/2$; and III) the cut-off scale $\omega_1(t)$, which controls the position of the so-called 'end point' of the spectrum, beyond which the production of gravitational radiation becomes exponentially small. We underline the fact that equation (3.26) is generally valid for all models with a dilaton.

If we add also the PBB phase, we can modify equation (3.26) and get the full energy spectrum for this class of minimal PBB models as:

$$
\Omega_{GW}(\omega, t_0) = \left(\frac{H_1}{M_{Pl}}\right)^2 \Omega_r(t_0)\left(\frac{\omega}{\omega_1}\right)^{3-2|1/2-\alpha|} \qquad \omega_s < \omega < \omega_1
\tag{3.27}
$$

$$
= \left(\frac{H_1}{M_{Pl}}\right)^2 \Omega_r(t_0)\left(\frac{\omega_s}{\omega_1}\right)^{-2|1/2-\alpha|}\left(\frac{\omega}{\omega_1}\right)^3 \qquad \omega_{eq} < \omega < \omega_s
\tag{3.28}
$$

where ω_s is the proper frequency of the transition scale. The interested reader is again advised to look at Gasperini's book [14] for calculations and further details.

The most striking difference with the conventional inflationary scenario is the fast growth of the spectrum with frequency. This is our most solid hope to detect such a spectrum with advanced ground-based detectors. The total energy density at ω_1, the cut-off frequency, depends on the ration between the string mass and the Planck mass:

$$
\Omega_{GW}(\omega_1, t_0) \simeq 8 \times 10^{-5}\left(\frac{M_s}{M_{Pl}}\right)^2
\tag{3.29}
$$

3.4.1 Massive dilatons

The dilaton field is an essential component of all superstring models, and thus of cosmological scenarios based on superstring theories. It is fair to say that the dilaton is the main difference between string models and general inflationary models based simply on GR equations. In the post-inflationary Universe we may expect that the dilaton tends to approach a stabilised configuration either under the action of a potential, which attracts it to a local minimum, giving it a mass in a process of spontaneous symmetry breaking, or simply as a consequence of the standard, radiation-dominated dynamics. However, depending on the mass acquired by the dilaton after inflation, a cosmic dilaton background could not only be produced, but survive until the present epoch and thus also be detectable today, at least in principle. We won't spend many words on that, because it is still highly speculative. We mention only the scalar-tensor generalisation of the standard geodetic deviation equation:

$$\frac{D^2\eta^\mu}{D\tau^2} + R^\mu_{\lambda\nu\rho}\eta^\lambda \dot{x}^\nu \dot{x}^\rho = q\eta^\nu \nabla_\nu \nabla^\mu \phi \tag{3.30}$$

where η^μ is a spacelike infinitesimal vector, q is a scalar charge, $R^\mu_{\lambda\nu\rho}$ is the Riemann tensor, D is the symbol of the covariant derivative and ϕ is the dilaton field. By using this equation and following the very same line of thought that led us to equation (2.67) we find a similar result for the SNR of the cosmic dilaton background:

$$\mathrm{SNR} = \frac{3NH_0^2}{8\pi^3}\left[2T\int_0^{+\infty}\frac{\mathrm{d}p}{p^3(p^2+m^2)^{3/2}}\frac{\gamma^2(p)\Omega_\chi^2(p)}{P_1\left(\sqrt{p^2+m^2}\right)P_2\left(\sqrt{p^2+m^2}\right)}\right]^{1/2} \tag{3.31}$$

where γ is the ORF for a scalar background, which does not coincide with the ORFs for the tensor waves, $\Omega_\chi(p)$ is the energy density of the dilaton background, P_1 and P_2 are the noise power spectra of the two chosen detectors. These results are valid for all types of detectors, not only interferometers, although currently only interferometric detectors are in operation.

References

[1] Losurdo G, Maggiore M, Matone G, Babusci D, Foffa S and Sturani R 2001 The stochastic gravitational-wave background *Gravitational Waves* ed I Ciufolini, V Gorini, U Moschella and P Fre (Boca Raton, FL: CRC Press) p 181

[2] Schwartz Matthew D 2014 *Quantum Field Theory and the Standard Model* (Cambridge: Cambridge University Press)

[3] Cottingham W N and Greenwood D A 2007 *An Introduction to the Standard Model of Particle Physics* (Cambridge: Cambridge University Press)

[4] James Rich 2001 *Fundamentals of Cosmology* (Berlin: Springer Verlag)

[5] Barbara Ryden 2003 *Introduction to Cosmology* (Boston, MA: Addison Wesley)

[6] Ramond Pierre 1999 *Journeys Beyond the Standard Model* (Cambridge, MA: Perseus Books)

[7] Durrer Ruth 2008 *The Cosmic Microwave Background* (Cambridge: Cambridge University Press)

[8] Weinberg Steven 2008 *Cosmology* (Oxford: Oxford University Press)

[9] Caldwell R R and Allen B 1992 Cosmological constraints on cosmic-string gravitational radiation *Phys. Rev.* D **45** 3447

[10] Birrell N D and Davies P C W 1982 *Quantum Field Theory in Curved Space* (Cambridge: Cambridge University Press)

[11] Ade P A R *et al* 2014 (BICEP2 Collaboration). Detection of b-mode polarization at degree angular scales by BICEP2 *Phys. Rev. Lett.* **112** 241101

[12] Adam R *et al* (The Planck Collaboration) 2017 planck intermediate results. XXX. The angular power spectrum of polarized dust emission at intermediate and high galactic latitudes *Astron. Astrophys.* **586** 121101

[13] Caprini C 2015 Stochastic background of gravitational waves from cosmological sources *J. Phys.: Conf. Ser.* **610** 012004

[14] Gasperini M 2007 *Elements of String Cosmology* (Cambridge: Cambridge University Press)

Chapter 4

Astrophysical sources

The first three confirmed detections of GWs, GW150914, GW151226 and GW170104, are all mergers of a black hole (BH) binary system. It is therefore likely that there are many more events of this kind, that we cannot resolve individually because they are too close in frequency. Those binary mergers would constitute a stochastic background of astrophysical origin, an astrophysical 'foreground'. Similarly we expect such a foreground also for neutron star (NS) binaries [1]. The meaning of the GW151914 detection for a stochastic background of binary black holes has been studied by the LIGO/Virgo Collaboration itself [2]. Recent estimates suggest that there are 10^8 NS in the Milky Way only, and that roughly 5% of all NSs live in NS/NS binary systems. A similar proportion might live in mixed NS/BH binaries, see [3] and references therein. Actually, for eLISA, such a foreground—of white dwarf binaries rather than BHs—is the main target source. It is expected [2] that such a foreground will not mask the cosmological background in the frequency range of ground-based detectors, although new results on it are expected. One of the most promising source of gravitational waves, potentially detectable by second and third generation detectors, is the stellar core collapse process. The physics of core collapse is complex and is expected to produce gravitational radiation through quite a few different mechanisms [4]. We mention briefly that there is a maximum frequency at which astrophysical sources can radiate. A source of mass M, even if very compact will be at least as large as its gravitational radius $2GM$, the bound being saturated only by BHs. Even if its surface were rotating at relativistic speed, its rotation period would be at least $4\pi GM$ and gravitational radiation cannot be produced with a period much shorter than that [5]. Therefore, we have a maximum frequency:

$$f \leq \frac{1}{4\pi GM} \sim 10^4 \frac{M_\odot}{M} \text{ Hz} \tag{4.1}$$

doi:10.1088/978-1-6817-4082-9ch4
4-1

As previously stated, a detection of a stochastic component of gravitational radiation in the frequency range spanned by ground-based detectors would, in all likelihood, come from the cosmological background rather than from the astrophysical foreground, whereas things are radically different for eLISA. A detection by eLISA of the astrophysical component would be of utmost importance for the GW community, as mentioned, such a foreground is eLISA's main source. At the same time it would be crucial if data analysis algorithms could be implemented to try to separate the supposedly anisotropic astrophysical foreground from the cosmological background.

4.1 Conclusions

That SGWB is by far the most difficult source of gravitational radiation to detect. At the same time it is the most interesting and intriguing one. Detecting it would mean, on the experimental side, that interferometric GW detectors were working even better that expected. On the observational side such a detection could give us information about the very early Universe, information that could not be obtained otherwise. Even negative results and improved upper bounds could put constraints on many cosmological and particle physics models. We kindly urge you, dear reader to stay tuned. The SGWB is worth the excitement and the waiting.

References

[1] Formichella V 2010 Onde gravitazionali da stelle di neutroni: valutazioni sulla rivelazione del fondo stocastico (Gravitational waves from neutron star binaries: an evaluation of the stochastic background, in Italian) *PhD Thesis* University of Turin

[2] Abbott B *et al* 2016 (LIGO scientific collaboration and virgo collaboration) GW150914: implications for the stochastic gravitational-wave background of binary black holes *Phys. Rev. Lett.* **115** 131102

[3] Lattimer James M and Maddapa Prakash 2007 Neutron star observations: Prognosis for equation of state constraints. *Phys. Rep.* **442** 109–65

[4] Crocker K, Mandic V, Regimbau T, Belczynski K, Gladysz W, Olive K, Prestegard T and Vangioni E 2015 Model of the stochastic gravitational-wave background due to core collapse to black holes *Phys. Rev.* D **92** 063005

[5] Losurdo G, Maggiore M, Matone G, Babusci D, Foffa S and Sturani R 2001 The stochastic gravitational-wave background *Gravitational waves* eds I Ciufolini, V Gorini, U Moschella and P Fre (Boca Raton, FL: CRC Press) p 181

www.ingramcontent.com/pod-product-compliance
Lightning Source LLC
Chambersburg PA
CBHW082113210326
41599CB00033B/6689